AN **INCONVENIENT** SEQUEL
TRUTH TO POWER

Your action handbook to learn the science, find your voice,
and help solve the climate crisis

AL GORE

RODALE.

MELCHER
MEDIA

Printed in the United States of America

Rodale Inc. makes every effort to use acid-free, mixed content recycled paper.

MIX
Paper from responsible sources
FSC® C132124

Library of Congress Cataloging-in-Publication Data is on file with publisher.

ISBN 978-1-63565-108-9 paperback

Distributed to the trade by Macmillan

2 4 6 8 10 9 7 5 3 1 paperback

Published by

We inspire health, healing, happiness, and love in the world. Starting with you.

733 Third Avenue
New York, NY 10017
rodalebooks.com

Produced by

124 West 13th Street
New York, NY 10011
melcher.com

This book is dedicated to my children:
Karenna Gore, Kristin Gore, Sarah Gore Maiani,
and Albert Gore III—and to my grandchildren,
Wyatt Schiff, Anna Schiff, Oscar Shiff, and Aria Maiani

At this point in the fight to solve the climate crisis, there are only three questions remaining:

Must we change?

Can we change?

Will we change?

Having spent the better part of my life for the past several decades trying to learn from experts on the climate crisis and working with technology and policy innovators to develop solutions for the unprecedented challenge humanity faces, I have never been more hopeful.

At this point in the fight to solve the climate crisis, there are only three questions remaining:

Must we change?
Can we change?
Will we change?

In the pages that follow, you will find the best available evidence supporting the overwhelming conclusion that the answer to the first two of these three questions is a resounding "Yes."

I am convinced that the answer to the third question—"Will we change?"—is also "Yes," but that conclusion, unlike the answer to the first two questions, is in the nature of a prediction. And in order for that prediction to come true, there must be a continued strengthening of the global consensus embodied in the Paris Agreement of December 2015, in which virtually every nation on Earth agreed to take concerted action to reduce net greenhouse gas emissions to zero as early in the second half of this century as possible.

That strengthened consensus depends in turn upon the continuing growth of a global grassroots movement to encourage political leaders in every nation to take even bolder steps than the ones agreed to in Paris. Luckily, that grassroots movement is already growing rapidly, not only among activists and leaders of civil society but also among business leaders, investors, mayors, and other elected officials who recognize that the stakes have never been higher. As more and more people come to the same realization that we really *must* change, the movement continues to gain momentum.

In other words, all three of these questions are intimately interrelated. In order to accelerate and complete the historic transformation of global civilization that is already under way, it is first necessary to come to grips with the unprecedented threat to humanity

A young woman holds a symbolic tulip at a climate change demonstration.

——

Paris, France
December 12, 2015

TRUTH TO POWER

I have never been
more hopeful.

Action is needed to prevent catastrophic harm.

posed by our continued reliance on fossil fuels, our unsustainable industrial transportation, agriculture, forestry, and ocean management practices—and our habit of short-term thinking that blinds too many of us to the unimaginable damage we are causing. And in order to summon the will to act with the requisite boldness, we have to have confidence that once we commit ourselves, we *can* succeed.

So let's begin with the first question: must we change?

In some ways, it is easy to understand one of the main reasons it has taken so much time to fully recognize the self-destructive nature of our current pattern. After all, humanity has gained immeasurable benefits from the burning of fossil fuels—higher standards of living, longer lifespans, historic reductions in poverty, and all of the blessings of the elaborate global civilization that has been developed over the past 150 years.

Moreover, because we still depend on fossil fuels for more than 80 percent of the energy that powers our civilization,

Arctic meltwater gushes from an ice cap.

——

Nordaustlandet, Norway
August 7, 2014

we are naturally daunted by the prospect of a rapid transition to renewable sources of energy and the speedy development of much higher levels of efficiency in all human activities.

Nevertheless, the obvious and overwhelming evidence of the damage we are causing is now increasingly impossible for reasonable people to ignore. It is widely known by now that there is a nearly unanimous view among all scientists authoring peer-reviewed articles related to the climate crisis that it threatens our future, that human activities are largely if not entirely responsible, and that action is needed urgently to prevent the catastrophic harm it is already starting to bring.

More importantly, Mother Nature is reminding us almost daily that the impacts of the climate crisis are growing steadily more severe, with more frequent and powerful climate-related extreme weather events. Every night, the TV news is like a nature hike through the Book of Revelation.

But before diving further into examples of the unprecedented harm we are causing, please remember how important it is to guard against feelings of despair. Despair, after all, is simply another form of denial, and can serve to paralyze the will we need to fight our way out of this crisis. And bear in mind that the hopeful news about the availability

Every night, the TV news is like a nature hike through the Book of Revelation.

of solutions is a powerful antidote to the feelings that can be aroused by the disconcerting news about the self-harm we are presently inflicting upon humanity.

High temperature records are being routinely broken on every continent—even Antarctica, where, in 2015, scientists confirmed one measurement of 17.5°C (63.5°F). Because of the physics of global warming, nighttime temperatures are rising even faster than daytime temperatures, and heat waves are becoming far more common.

Air temperatures are predicted to steadily increase all around the world because of the continuing accumulation of man-made global warming pollution in the atmosphere, the thin shell of air that surrounds our planet. The cumulative amount of this gaseous pollution exceeded 400 parts per million for the entire year for the first time in human history in 2016 (43 percent higher than preindustrial levels).

The combination of higher ocean temperatures and the growing acidification of ocean water (approximately one-third of the CO_2 we release into the atmosphere settles into the ocean, where it has already increased acidity by 30 percent) are leading to the death of coral reefs throughout the world. It is also disrupting the process by which coral polyps—and all sea creatures with shells—scavenge calcium carbonate from seawater and transform it into the hard structures necessary for their survival.

Warmer ocean temperatures are also causing the mass migration of fish populations, many of which are simultaneously being depleted by overfishing, the runoff of pollution from coastal areas, and a sharp decline of oxygen in the growing number of "dead zones" in the ocean. The decline of fish populations in the tropics and subtropics is especially threatening because of the heavy reliance on seafood in those regions.

The climate crisis is resulting in more

Day 21 after Hurricane Katrina.
—
Plaquemines Parish, Louisiana
September 19, 2005

TRUTH TO POWER

frequent downpours around the world, but the geographic distribution and the periodicity of rainfall have also been altered. As the water cycle is disrupted, some areas are receiving big increases in precipitation, while others are receiving much less. While much more precipitation falls in big storm events, the period of time between rainfalls has also been increasing in many regions. And during the intervals between rainfalls

the higher air temperatures suck more moisture from the first several centimeters of the soil, leading to deeper and longer droughts.

Driven by these droughts and warmer temperatures, there has been a dramatic increase in fires and a radical extension of the "fire season" (in the western United States, for example, the annual fire season has already increased by 105 days). Firefighters are now having

Firefighters are now having to deal with a new phenomenon they call "mega-fires."

to deal with a new phenomenon they call "mega-fires." These extremely large fires are particularly severe in the northern boreal forests of Alaska, Canada, and Siberia. The higher temperatures have also greatly increased the damage to forests from pine beetles and bark beetles, which survive milder winters in greater numbers and reproduce more generations during the longer summers.

When they attack trees weakened by drought, they end up devastating millions of acres of forestland, turning them into kindling for the larger fires.

The seasonal timing of rainfall patterns has become more erratic and less predictable, which has had a harsh impact on subsistence farmers who depend on rain-fed agriculture. Especially in the tropics and subtropics, many farmers report that they can no longer rely on the age-old rainy season/ dry season pattern that previous generations counted on to decide when to plant and when to harvest.

Because the larger downpours rush off the surface of the land, carrying larger amounts of topsoil with them, the underground aquifers beneath are not being replenished as efficiently as they are by more regular, gentler rains. And because higher temperatures increase the need for water—by plants and animals, in agriculture and energy production, and for human consumption—growing populations in many regions have begun to rely more heavily on underground aquifers, depleting those sources more rapidly than they are replenished.

Increasing temperatures are also accelerating the melting and breakup of land-based ice—in mountain glaciers, and more significantly, in Greenland and Antarctica—thus accelerating sea level rise and threatening the inundation of low-lying coastal areas where hundreds of millions of people live.

Food supplies are also being threatened by the climate crisis. Higher temperatures are already decreasing crop yields of corn, wheat, rice, and other staples.

Diseases affecting humans have become more threatening because of the climate crisis. The so-called "vectors" that carry many infectious diseases—such as mosquitoes, ticks, snails, algae, and others—are extending their range in a warmer and wetter world. Viruses such as Zika, dengue fever, and West Nile virus, among others, incubate faster.

Climate has always mediated the relationship between human beings and microbes. Indeed, some historians have suggested that civilizations in higher latitudes have more readily flourished because they are relatively freer from the overburden of microbial diseases that are much more prevalent in the hotter and wetter regions of the lower latitudes of the tropics and subtropics. According to Princeton's Andrew Dobson, "Climate change is disrupting natural ecosystems in a way that is making life better for infectious diseases."

One of the most startling examples of the new threat from tropical diseases is the recent outbreak of a mutated form of the Zika virus that causes microcephaly and other serious birth defects. Zika is also the first mosquito-borne disease known to be sexually transmitted by humans. Although air travel has played a large role in the spread of tropical diseases, the changing climate conditions have given them a chance to take root in regions where they never thrived previously.

For all of these reasons, and more, the answer to the question "Must we change?" is abundantly clear: "Yes!" Indeed, some scientists have warned that if we do not change, the future of human civilization itself is at dire risk.

But here is the good news: the answer to the second question—"Can we change?"—is also a resounding "Yes!" In just the past 10 years the cost of clean, renewable solar energy and wind energy has fallen so quickly that in a growing number of regions throughout the world, it is now cheaper than electricity made from the burning of fossil fuels—and the cost continues to decline year by year.

Some areas of technology—such as computer chips, flat screen televisions, and mobile telephones—respond to research and development spending in an almost magical manner. Not only does the cost go down much faster than anyone expected, the quality goes up at the same time.

Consider, for example, what has happened with cell phones: in 1980, AT&T took note of the new technology that led to the first bulky mobile telephones and asked one of the world's leading market research companies, McKinsey, to predict for them how many such mobile phones they might be able to sell by the year 2000. They were excited to get the answer that they could probably count on selling 900,000 such phones. When the year 1999 arrived, the telephone industry did sell 900,000 mobile phones—in the first three days of the year! By the end of the year, 109 million phones had been sold—120 times more than predicted.

Smoke rises from a factory in Chiba province, a hub for Japanese industry.

—

Soga City, Japan
February 14, 2008

TRUTH TO POWER

So the answer to the question "Must we change?" is abundantly clear: "Yes!"

The cost had come down far more quickly than anyone anticipated, even as the quality of the phones increased dramatically, with many more features packed into a much smaller form factor. Moreover, these cell phones proved especially popular in developing countries, where landline telephone grids had never been built in the first place. The people in those regions simply leapfrogged the technology on which the wealthy countries were still dependent.

The same thing is now happening with renewable energy, especially solar. Panels are being installed on the roofs of grass huts in Africa and South Asia— and on rooftops throughout the world.

And here is the good news: the answer to the question "Can we change?" is also a resounding "Yes!"

Utility-scale "solar farms" are producing electricity at even lower rates. In fact, some of the new contracts for large solar farms signed in 2016 provide electricity at rates less than half the cost of the cheapest electricity generated from the burning of coal or natural gas. And the price continues to decline rapidly.

As a result, just as the early predictions of growth in the number of cell phones were badly wrong, so were the best predictions made 15 years ago on the spread of solar and wind energy.

For the year 2016 in the United States, 70 percent of all new electricity-generating capacity came from solar and wind. Less than two-tenths of 1 percent came from coal. And in a growing number of areas, solar electricity is now beating proposals to generate electricity from the most efficient forms of gas turbines.

To brighten the renewable energy story still further, the cost of battery storage is also now beginning to rapidly decline. This is particularly important, because more efficient, cheaper batteries can solve the so-called "intermittency" disadvantage of renewable sources—that is, they can continue providing electricity at nighttime when the sun doesn't shine and during periods of the day when the wind is slack. As Noah Smith, a *Bloomberg View* columnist, wrote, "Solar-plus-batteries is set to begin a dramatic transformation of human civilization."

Non-polluting electric vehicles are also beginning to make inroads in the transportation sector of the economy. While their percentage of the vehicle fleet is still small, demand is growing rapidly and competition is driving virtually every major vehicle manufacturer to introduce more affordable models in the next 12 months. The leading electric vehicle manufacturer in the United States, Tesla, has just surpassed General Motors and Ford to become the most valuable car company in the nation. Homeowners with solar panels on their rooftops and an electric car in their garages are already paying less than the equivalent of $1 per gallon in gasoline costs. Electric buses are predicted to take over from diesel buses in major cities over the next decade.

Moreover, by using the powerful new digital tools now available in our civilization—including the fast-growing "Internet of Things"—industrial

This Japanese solar power plant will supply energy to approximately 22,000 homes.

Kagoshima, Japan
August 13, 2015

The fossil fuel industry is engaged in a losing battle to confuse people into thinking that the climate crisis isn't real, and that the renewable energy revolution is trivial and meaningless.

Smoke billows from a coal plant at sunset.

—

Portugal

and business managers are achieving much higher levels of efficiency in the use of energy and natural materials. Thousands of new technology solutions are spreading rapidly through the global economy. LED lighting is displacing incandescent bulbs and compact fluorescent light bulbs at an unprecedented rate in what is now being called perhaps the fastest technology transformation in any market sector in the history of the world.

In fact, taken together, the spread of renewable energy, battery storage, electric vehicles, LEDs, and the thousands of new hyper-efficiency solutions all make up what many are now referring to as "The Sustainability Revolution." It combines the *scale* of the Industrial Revolution with the *speed* of the Digital Revolution.

The fossil fuel industry, which is already engaged in a losing, rearguard battle to confuse people into thinking that the climate crisis isn't real, is also engaging in what some have described as "a strange new form of denial"—an effort to convince people that the renewable energy revolution is trivial and meaningless. In their failing efforts to both persuade people the crisis isn't real and that solar and wind energy are not cost-effective solutions to it, they bring to mind the famous scene in the Marx Brothers movie *Duck Soup*, in which

Workers walk past the vast cluster of solar panels at the Gujarat Solar Park.

———

Gujarat, India
April 14, 2012

Chico asked, "Who you gonna believe, me or your own eyes?"

Importantly, investors around the world are ignoring the fossil fuel industry's arguments and are now shifting resources massively into renewable energy and are financing the highly profitable Sustainability Revolution. As the renewable energy industry has flourished, the market capitalization of the global coal industry has fallen almost *90 percent* in the past seven years.

And with the International Energy Agency warning that two-thirds of the proven reserves held by oil and gas companies can never be burned—lest human civilization be destroyed—investors are beginning to realize that they are in danger of the kind of financial catastrophe that was caused in 2007–2008, when the value of subprime mortgages was suddenly discovered to be worthless. In the same way, "subprime carbon assets" now pose a serious threat to the stability of the global economy. And smart investors are moving quickly to minimize their exposure to this historic risk.

The Sustainability Revolution combines the *scale* of the Industrial Revolution with the *speed* of the Digital Revolution.

It is also important to note that the conventional air pollutants from the burning of coal and other fossil fuels that accompany CO_2 emissions—including particulates and sulfur dioxide—have now reached intolerable levels in many cities around the world—especially in China and India—where people are becoming desperate to reduce the levels of pollution. The health problems associated with air pollution are driving political unrest in some areas as people become more aware of the dangers they face.

So the answer to the second question—"Can we change?"—is clearly "Yes."

But that leaves the third and final

I have heard over and over again a hunger to engage in this struggle for the future. That gives me hope every day.

question: "Will we change?"

I am convinced that the answer to this question is also "Yes." The global agreement reached in Paris at the end of 2015 is encouraging not only because of the universal commitment from governments but also because it amplified the signal to investors, industries, businesses, and institutions that the entire world is poised to move quickly to a sustainable and renewable future. This transformation has already begun and is picking up speed. The Paris Agreement also calls for regular five-year reviews of the commitments made by the nations that are party to the agreement to encourage them to increase their commitments to emission reductions as the new technologies become ever cheaper and as opposition to the pollution from fossil fuels grows ever stronger.

Many state, provincial, regional, and local governments are already moving faster than national governments. A growing number of cities in the United States have succeeded in obtaining 100 percent of their electricity from renewable sources, and many regions around the world are making similar progress.

While the answer to "Will we change?" is almost certainly "Yes," it is not yet clear that we will change rapidly enough to avoid the catastrophic damage we *must* avoid. Yet it is abundantly clear that we *can* avoid it if we accelerate the pace of transformation.

And that is why, starting on page 176, you will find an action handbook for anyone who wants to be a part of the answer to the biggest question our civilization has ever confronted.

Because many governments in the world—especially the United States government in 2017—are still controlled by fossil fuel interests, the growing citizen activist movement pushing for more rapid change is actually the most important movement in world history.

As I have traveled the world, including during the filming of *An Inconvenient Sequel: Truth to Power*, the companion film to this book, I have heard over and over again a hunger on the part of everyday citizens to engage in this struggle for the future of our civilization. And that gives me hope every day.

If you want to be a part of this historic change, you will find in the following pages not only a more detailed description of the crisis and its solutions but also a list of actions that you can take to join the climate movement and ensure that we *will* change—and that we will change in time. ◉

If we were to fail, the next generation would be well justified in looking back at us and asking:

What were you thinking?

Couldn't you hear what the scientists were saying?

Couldn't you hear what Mother Nature was screaming at you?

The original Rapa Nui people of Easter Island, who created these statues, nearly disappeared due to the exhaustion of natural resources in the area.

Easter Island, Chile
January 1998

There's a hunger for information about *what's happening,* *why* it's happening, and *how* we can fix it. **Ten years ago, I launched a training program, the Climate Reality Project, for anyone who wanted to learn how to communicate to others the threat of and solutions to the climate crisis. What follows are selections from that training program.**

O POWER

I always start my trainings by showing this photo, *Earthrise.* This was the first picture that humans saw of the Earth taken from space. It had a profound impact on the consciousness of humanity. This was the first time that human beings left near-Earth orbit and went far enough into space to see the planet whole, floating in the void.

Earth, with the lunar landscape in the foreground.

—

Apollo 8 mission
December 24, 1968

Our perspective on the ground looking up at the sky can mislead us. We see the atmosphere as a vast and limitless expanse. But scientists have long known what this picture from the International Space Station shows: the sky is actually an extremely thin shell surrounding our planet. That is what makes it so vulnerable to human activities.

If you could drive straight up in a car at highway speeds, you'd be in outer space in under an hour.

We put 110 million tons of global warming pollution into this very thin space every day.

Earth's atmosphere, from the International Space Station.

—

June 2014

One reason we've failed to recognize the damage we're doing is that we've assumed it's fine to use our atmosphere as an open sewer.

The basic science of global warming is straightforward, but it is important to understand it fully.

Energy from the sun enters the Earth's atmosphere as light. The wavelength of light is very short, so it cuts right through the atmosphere and it warms the planet. Then, that heat energy is radiated back into space in the form of infrared radiation, which has a longer wavelength. Some of that outgoing infrared radiation is trapped by the natural layer of greenhouse gases, which has long been just the right thickness for supporting life. It keeps the temperatures on Earth within a comfortable range: not too hot like Venus, and not too cold like Mars.

The Earth's natural greenhouse gas layer is now being transformed, mainly by the burning of fossil fuels.

This pollution makes the layer thicker, which traps more outgoing infrared radiation, also known as heat.

1. Energy in the form of light from the sun enters the atmosphere and is absorbed by the Earth, warming it.

2. Some energy is radiated back into space in the form of infrared waves.

3. When greenhouse gases build up in the atmosphere, more infrared radiation is trapped. This makes the planet grow hotter.

In 2006, when *An Inconvenient Truth* was released, the concentration of CO_2 in the atmosphere had reached 382 parts per million—up from a preindustrial level of 280 ppm.

In March 2017, the level reached an all-time high of 409.5 ppm.

Many scientists believe that a "safe" level for humanity is 350 ppm.

CARBON DIOXIDE CONCENTRATION MEASURED AT MAUNA LOA OBSERVATORY, HAWAII

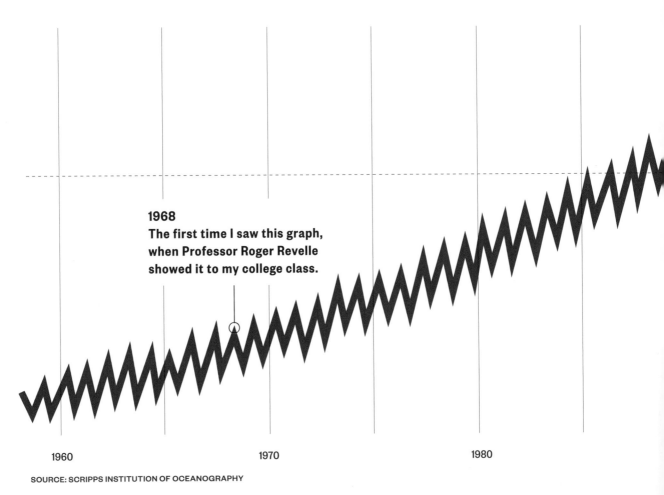

1968
The first time I saw this graph, when Professor Roger Revelle showed it to my college class.

1960

1970

1980

SOURCE: SCRIPPS INSTITUTION OF OCEANOGRAPHY

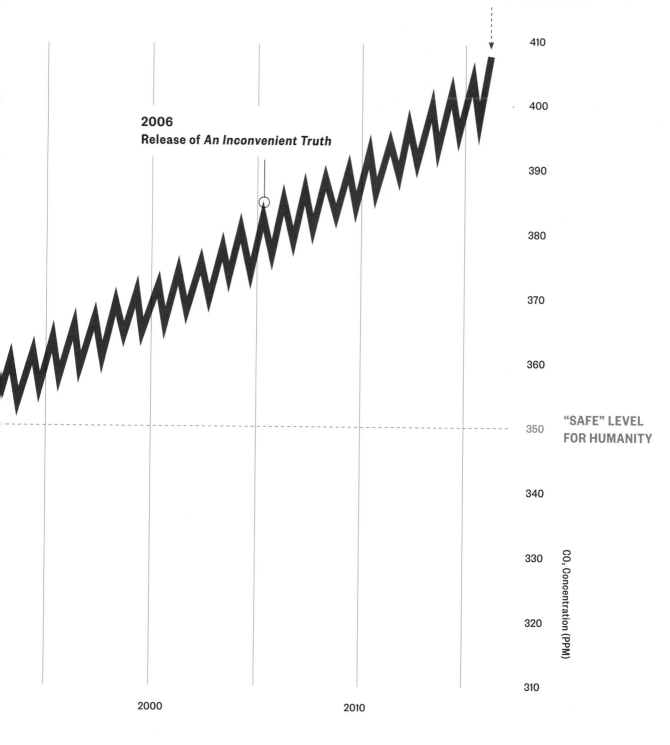

ALL-TIME HIGH AS OF MARCH 2017

2006
Release of *An Inconvenient Truth*

"SAFE" LEVEL
FOR HUMANITY

CO₂ Concentration (PPM)

410

400

390

380

370

360

350

340

330

320

310

2000

2010

Common sources of greenhouse gases include agriculture, mining, and the burning of forests. But the principal source is the burning of carbon-based fuels. After World War II, the use of fossil fuels began to increase dramatically. In recent years, it has accelerated even more.

This is the heart of the problem. If we solve this, it will be much easier to address the other sources.

THAWING PERMAFROST

COAL BURNING

COAL MINING

INDUSTRIAL PROCESSES

AIR TRANSPORTATION

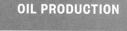

OIL PRODUCTION

CROP BURNING

FOREST BURNING

FRACKING

NITROGEN FERTILIZER

LAND TRANSPORTATION

LANDFILLS

INDUSTRIAL AGRICULTURE

Why Do the Scientists Feel So Strongly?

AS THE CLIMATE CHANGES, so do a variety of wind and ocean patterns, along with the distribution of plant and animal life. That's why the U.S. Environmental Protection Agency (EPA) has tracked 37 different indicators of climate change on land and over water, including surface temperature, bird wintering ranges, and coastal flooding. Taken together, the data collected paints a picture of our climate and its changing impact on all of Earth's systems. But where does this data come from? What instruments are used? In short, how do we really know what's happening to the climate now—and what happened to it in the past?

Ice cores are a good place to start. These long cylinders of ice are drilled out of the glaciers and ice caps covering Greenland and Antarctica. The lowest layers of them date back thousands of years. In fact, the oldest ice core records, from Antarctica, extend back 800,000 years. The U.S. National Ice Core Laboratory is home to more than 17,000 meters of ice from all over the world that can be used to understand the planet's climate history. These layers of ice store tiny bubbles of air that were trapped by the snow when it fell. By measuring the ratio of different isotopes of oxygen in these air bubbles, scientists can re-create both the CO_2 content of the air year by year and the temperature of the air that was trapped.

While ice cores enable scientists to build the record of past millennia, sophisticated, modern tools also allow them to better understand present and future climate patterns. NASA operates a fleet of satellites that collect a broad range of climate information. For example, the GRACE (Gravity Recovery and Climate Experiment) missions record the steady retreat of ice sheets in astonishing detail, while the Jason-3, OSTM/Jason-2, and Jason-1 missions track sea level rise, which has increased by an average of three inches since 1992. You can read about another satellite, DSCOVR, on page 164. The National Oceanic and Atmospheric Administration operates sophisticated data collection instruments, including underwater drones, buoys, and shipborne instruments, to measure the oceans' carbon levels and temperatures in order to better predict the impacts of climate change. Other devices used by scientists include weather balloons, ships, and radar, each of which contribute their own unique types of data to create a big picture of our climate at any given time. ◉

Scientists use ice cores to measure levels of atmospheric carbon and other gases trapped in layers of ice. Ross Sea, Antarctica, November 16, 2011

Katharine Hayhoe

Codirector of Texas Tech University's Climate Science Center

LUBBOCK, TEXAS

KATHARINE HAYHOE HAD BEEN married for six months before she realized that her husband didn't think climate change was a scientific fact.

"I had no idea that there were people who didn't agree with the science," said Hayhoe, who is the codirector of Texas Tech University's Climate Science Center.

The realization led the couple to embark on a grueling debate that would last years. It was ironic, in retrospect, because Hayhoe was already a rising star in climate modeling, working at the University of Illinois Urbana-Champaign under famed researcher Donald Wuebbles in the emerging field of statistical downscaling.

Statistical downscaling combines historical weather observations with computer models to predict how climate change will affect a particular geographic region. It's a field of study that offers intimate portraits of the climate impacts on particular communities, which has led Hayhoe to approach her work with great empathy.

"If you discover that your patient has a new, very serious disorder, you don't just write a paper about it," she said. "You try to help them. You can't say, 'Oh, that's not really my job.'"

Hayhoe has used her computer models to consult on the effects of climate change for the cities of Washington, D.C., Boulder, Colorado, and Chicago, as well as for the Department of Defense and the United States Fish and Wildlife Service. Her recommendations can cover topics ranging from urban planning considerations like managing sewers to how storms and droughts will affect agricultural infrastructure.

But the worlds of academia and policy aren't enough for Hayhoe, who also goes out of her way to connect with climate change deniers, often on their home turf. In those ad-hoc debates, which have played out at symposiums and city council meetings across the nation, Hayhoe remains unflappable, and never misses a chance to build bridges to those she speaks with.

And in the end, Hayhoe won the debate with her husband, who now teaches linguistics at Texas Tech and serves as the pastor of a small church. In fact, the two ended up coauthoring a book titled *A Climate for Change: Global Warming Facts for Faith-Based Decisions*.

"We were highly motivated to work it out," Hayhoe said dryly. "We were married, and we wanted to stay that way." ◉

See page 226 for more information about talking with climate deniers.

We are now trapping as much extra heat energy in the atmosphere as would be released by

400,000 Hiroshima-class atomic bombs exploding on the Earth's surface every day.

We live on a big planet, but that is an unimaginable amount of heat energy.

The mushroom cloud from a Hiroshima-class explosion in Operation Upshot-Knothole.

Nye County, Nevada
May 25, 1953

45

It's getting much warmer very quickly. This classic bell curve shows the distribution of summer air temperatures. Scientists use the 30 years between 1951 and 1980 as a so-called normal period, when there were roughly equal number of days with average temperatures, cooler than average temperatures, and warmer than average temperatures. In the 1980s, the entire curve shifted toward the warm side. And for the first time there were a statistically high number of extremely hot days, showing at the lower right in red. Then

SHIFT IN SUMMER TEMPERATURES, 1951–2015

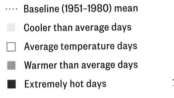

- ···· Baseline (1951–1980) mean
- ▢ Cooler than average days
- ☐ Average temperature days
- ▨ Warmer than average days
- ■ Extremely hot days

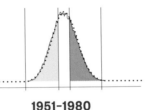

1951–1980

1983–1993

1994–2004

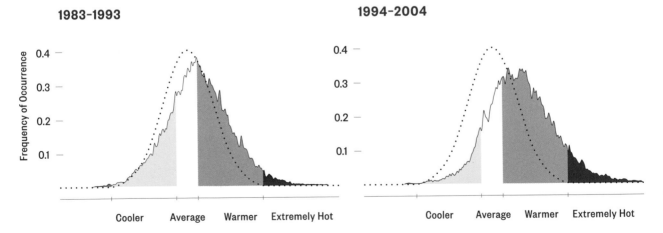

SOURCE: HANSEN, ET AL., 2016

in the 1990s, the curve shifted farther to the warm side and a combination of warmer and extremely hot days began to dominate. Over the past 10 years, the extremely hot days have become more numerous than the cooler than average days. In fact, the extremely hot

days are now almost 150 times more common than they were just 30 years ago.

There are still cold days, but they are far less frequent.

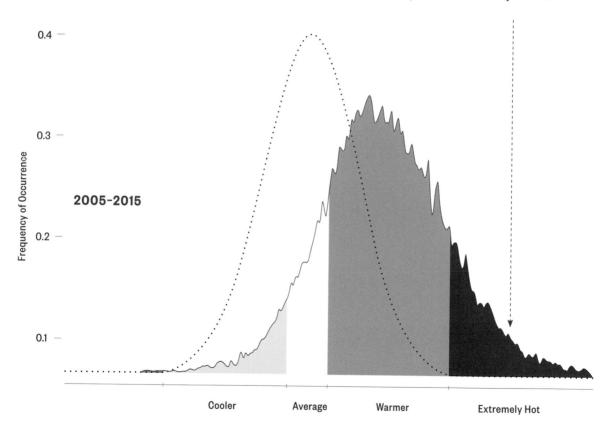

Extremely hot days used to cover 0.1 percent of the Earth. They now cover 14.5 percent.

When you combine high temperatures with high humidity, it feels even hotter. The "heat index" is a measure of how hot it actually feels to our bodies.

In July 2015, the heat index in Bandar Mahshahr, Iran, reached 74°C (165°F).

Iranians cool off at the Cheshmeh-Ali pool during a heat wave that caused temperatures to exceed 40°C (104°F).

—

Shahre-Ray, Iran
July 22, 2016

"The climate in large parts of the Middle East and North Africa could ... render some regions uninhabitable which will surely contribute to the pressure to migrate."

—Jos Lelieveld, director, Max Planck Institute for Chemistry

On his late night show, Jimmy Kimmel said, "You know how you know climate change is real? When the hottest year on record is whatever year it currently is."

In fact, 16 of the 17 hottest years ever measured with instruments (a practice that dates back to 1880) have occurred in the past 17 years. And the hottest year of all was 2016. The second hottest was the year before, and the third hottest was the year before that.

GLOBAL LAND AND OCEAN TEMPERATURE ANOMALIES, 1880–2016

Compared to average temperature 1901–2000

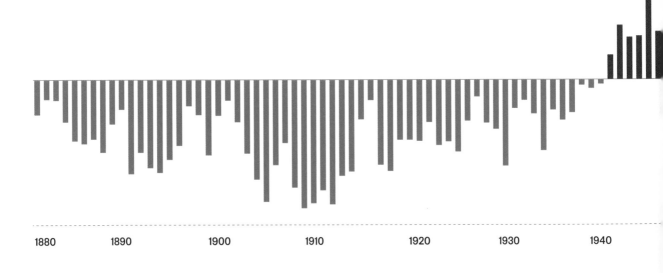

1880   1890   1900   1910   1920   1930   1940

SOURCE: NOAA

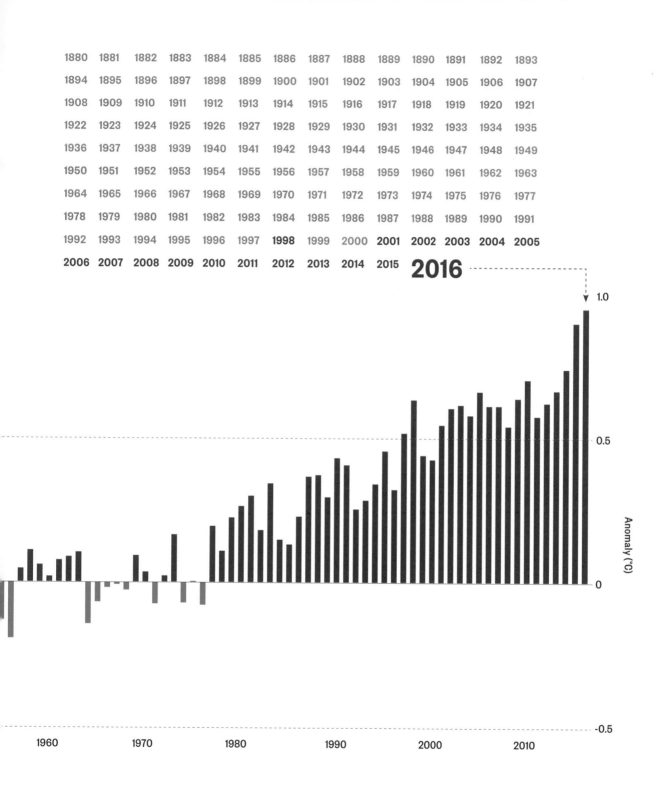

1880 1881 1882 1883 1884 1885 1886 1887 1888 1889 1890 1891 1892 1893
1894 1895 1896 1897 1898 1899 1900 1901 1902 1903 1904 1905 1906 1907
1908 1909 1910 1911 1912 1913 1914 1915 1916 1917 1918 1919 1920 1921
1922 1923 1924 1925 1926 1927 1928 1929 1930 1931 1932 1933 1934 1935
1936 1937 1938 1939 1940 1941 1942 1943 1944 1945 1946 1947 1948 1949
1950 1951 1952 1953 1954 1955 1956 1957 1958 1959 1960 1961 1962 1963
1964 1965 1966 1967 1968 1969 1970 1971 1972 1973 1974 1975 1976 1977
1978 1979 1980 1981 1982 1983 1984 1985 1986 1987 1988 1989 1990 1991
1992 1993 1994 1995 1996 1997 **1998** 1999 2000 **2001 2002 2003 2004 2005**
2006 2007 2008 2009 2010 2011 2012 2013 2014 2015 2016

1.0

0.5

Anomaly (°C)

0

1960   1970   1980   1990   2000   2010

-0.5

The streets are melting. We have built a civilization for conditions that we are now in the process of radically changing.

A road melt caused by temperatures exceeding 45°C (113°F).

———

New Delhi, India
May 24, 2015

In 2003, a European heat wave centered in France killed 70,000 people.

In 2015, a heat wave in Pakistan killed more than 2,000 people.

In the same summer, a heat wave in India killed at least 2,500 people.

A Pakistani man waits while volunteers search for the body of a deceased relative at the Edhi Foundation morgue.

Karachi, Pakistan
June 22, 2015

More than 90 percent of all the heat energy trapped by man-made global warming pollution goes into the ocean.

As a result, the buildup of ocean heat content has increased dramatically—especially in the past 25 years—and it is accelerating.

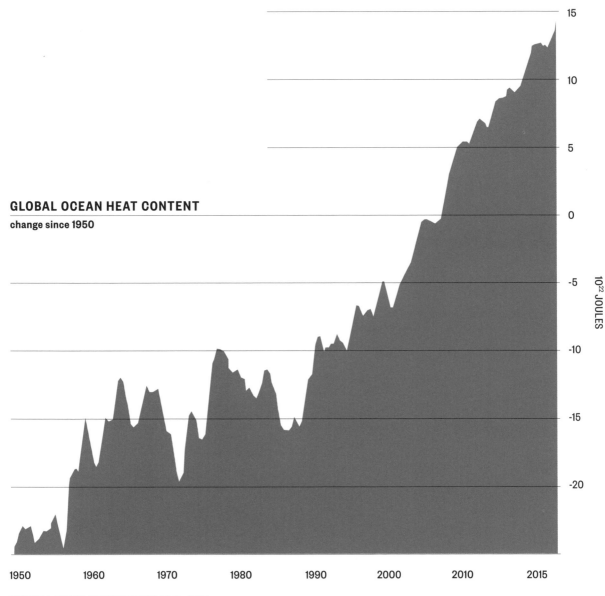

GLOBAL OCEAN HEAT CONTENT
change since 1950

10²² JOULES

15
10
5
0
-5
-10
-15
-20

1950   1960   1970   1980   1990   2000   2010   2015

SOURCE: L. CHENG, K.E. TRENBERTH, ET AL., 2017

A direct consequence is that when ocean-based storms cross warmer ocean waters, they pick up more convection energy. The storms get stronger and more destructive.

Super Typhoon Haiyan (known as Yolanda in the Philippines) crossed areas of the Pacific 3°C (5.4°F) warmer than normal. When it struck the city of Tacloban in the Philippines, it had become the strongest ocean-based storm to ever make landfall.

PATH OF TYPHOON HAIYAN

+ 1°C (1.8°F) Sea surface temperature anomaly + 5°C (9°F)

PHILIPPINES

TYPHOON HAIYAN

SOURCE: JAPAN METEOROLOGICAL AGENCY VIA WEATHER UNDERGROUND

A survivor stands among the wreckage left behind by Super Typhoon Haiyan.

Tacloban City, Philippines
November 10, 2013

Super Typhoon Haiyan caused 4.1 million refugees, some of whom are still not back in their homes. I traveled to Tacloban and visited some of the graves of the thousands of victims. Families were invited to pick one of these graves and inscribe the name of their loved one on it.

Visiting the Holy Cross cemetery, which houses the remains of thousands of Typhoon Haiyan victims.

Tacloban City, Philippines
March 11, 2016

John Leonard Chan

Climate Reality Leader

TACLOBAN CITY, PHILIPPINES

JOHN LEONARD CHAN recalled the morning when the typhoon began. The winds started to tear at his family's single-story dwelling in Tacloban, a small city on the Philippine island of Leyte.

By midmorning, floodwaters were rushing into the home. But transportation and communication systems had broken down during Typhoon Haiyan, so the family stayed put for two terrifying days before setting off on foot for Chan's grandparents' house. The journey took three hours of climbing over debris, and Chan remembers seeing dead bodies in the street.

"People were crying, people were screaming," said Chan, now 22, of the chaos after the 2013 typhoon. "Nobody knew what to do."

The family eventually secured enough food to survive for five days until they were evacuated to Manila, where they have lived ever since.

As his family adjusted to life in Manila, Chan became increasingly concerned with how climate change might have given rise to Haiyan, one of the most powerful cyclones in the history of the world.

"That's what made me interested in giving voice that climate change is real," he said. "I'm a first-hand survivor of an effect of climate change. I think my voice will be interesting to others."

In 2016, I met Chan at the Climate Reality Project's three-day training session in Manila, where he became a climate leader. Since then, he has also joined the Philippine Youth Climate Movement, a student group that raises awareness about climate change at home and abroad, and has given seminars about the science of global warming at high schools in Manila.

Chan is also leveraging his education into opportunities to more deeply study the effects of climate change. He's currently enrolled in a microbiology program at the University of the Philippines Los Baños, and last year he returned to the Tacloban area to study how the typhoon has impacted ocean life in the region.

Chan isn't sure yet what he intends to do after graduate school. But he's sure of one thing: he's going to continue to do everything in his power to give voice to the victims of climate change.

"There are people suffering because of climate change," he said. "People around the world need to hear from people from Tacloban, so they can understand that climate change is real." ◉

Turn the page to learn more about the Climate Reality Project.

What Is The Climate Reality Project?

Climate Reality: How It Began

When *An Inconvenient Truth* was released back in 2006, no one knew what to expect. I hoped it would spark a conversation but certainly didn't expect the incredible response and enthusiasm from people across the globe wanting to take action to solve the climate crisis.

These were people who'd probably never thought of themselves as activists, but now that they understood what the climate crisis meant for the planet and the people they loved, they wanted to make a difference.

So I invited 50 people to come to my farm in Tennessee and trained them—in the barn—to use the slideshow in the movie and talk to others about the crisis and how to solve it.

Climate Reality: Its Work Today

What began on that day in the barn has grown to become The Climate Reality Project, an international nonprofit with headquarters in the U.S. and branches in Australia, Brazil, Canada, China, Europe, India, Indonesia, Mexico, the Philippines, and Africa.

Our mission is daunting but simple: catalyze a global solution to the climate crisis by making urgent action a necessity across every level of society.

We've grown and changed a lot since those early days. But what hasn't changed is our commitment to training and empowering regular people to change the world. One conversation at a time.

Now we train thousands each year in multiday events, not in barns but in cities from Denver to Delhi. The people who come as concerned citizens leave as world changers we call Climate Reality

Leaders. Today, this global network of more than 12,000 activists (and counting) is out there spreading the word and organizing friends, neighbors, and communities for climate solutions in 136 countries.

Of course, today, the conversation online is also important. With social media a critical battlefield in the fight not just for hearts and minds, but for the very idea of truth—upon which democracy is founded—we give people the digital tools to combat climate denial, share good news about solutions, and put real pressure on policy makers.

The story doesn't end there. Our 100% Committed campaign enlists citizens to help cities, colleges, and businesses switch to 100 percent renewable electricity. We hold Days of Action where people take over the streets to build public support for key policies that accelerate the switch to clean energy and protect our climate.

And because we approach our work as I've always done—speaking to people where they're at and speaking truth to power—we're building a

The Climate Reality Project logo, which includes the iconic *Blue Marble* photo.

movement of activists capable of incredible things and winning against the odds around the world.

We saw it in Florida in 2016, when the fossil fuel industry tried to kill solar with a misleading ballot initiative. Along with others, we shared the truth about how solar could cut costs and create jobs. Voters responded, defeating the initiative and winning a resounding victory for Florida and our planet.

We're seeing it on college campuses around the world, where young student activists are pushing their schools to choose a clean energy future and shift to 100 percent renewable electricity through our 100% Committed campaign. Colorado State University, Plymouth State University, Hampshire College, and the University of Wisconsin-Stevens Point have already made the commitment and more are on the way.

This is only part of our story and it's only beginning. You can be a part of the next chapter. So if you're ready to make a difference, visit *climaterealityproject.org* to join Climate Reality and we'll show you how. ◉

Climate Reality Leadership Corps

WHAT: Trains citizens to become powerful change agents.

MAKING A DIFFERENCE: Climate Reality Leaders led successful campaigns to push Brazil to ratify the Paris Agreement and protect clean energy in Florida.

GET INVOLVED: climaterealityproject.org/training

Climate Speakers Network

WHAT: Trains community leaders as climate storytellers.

MAKING A DIFFERENCE: Thousands of faith, business, health, Latino, and African-American leaders become trusted messengers.

GET INVOLVED: climatespeakers.org

24 Hours of Reality

WHAT: 24-hour global broadcast highlighting the crisis and how we solve it for millions.

MAKING A DIFFERENCE: 30 million live views online in 187 nations (2016).

GET INVOLVED: 24hoursofreality.org

100% Committed

WHAT: Organizes cities, college campuses, businesses, ski resorts, and institutions to switch to 100 percent renewable electricity.

MAKING A DIFFERENCE: Salt Lake City, Park City, and Moab in Utah have committed to 100 percent renewable electricity.

GET INVOLVED: climaterealityproject.org/content /roadmap-100

Digital Action

WHAT: Digital tools to raise awareness of climate solutions and pressure lawmakers to act.

MAKING A DIFFERENCE: United over 2.2 million people in calling for action at COP 21, helping create overwhelming public pressure leading to the historic Paris Agreement.

GET INVOLVED: climaterealityproject.org/joinreality; facebook.com/climatereality; twitter.com/ climatereality

When *An Inconvenient Truth* came out, the most criticized scene was an animation showing that the combination of sea level rise and storm surge would put ocean water into the 9/11 Memorial site, which was then under construction. Some climate deniers said that was ridiculous, an exaggeration.

Superstorm Sandy crossed areas of the Atlantic 5°C (9°F) warmer than normal and became a monstrous storm.

BOTTOM: Projected flooding of the 9/11 Memorial site from *An Inconvenient Truth* (2006).

TOP: Actual flooding of the Memorial site from Superstorm Sandy (2012).

○ World Trade Center site
■ Actual Flooding

A satellite image of Superstorm
Sandy off the east coast of the
United States.

___

Atlantic Ocean
October 29, 2015

On the night of October 29th, 2012, seawater flooded the 9/11 Memorial site.

Seawater floods the 9/11 Memorial construction site in New York during Superstorm Sandy.

—

New York, New York
October 29, 2012

Months after Superstorm Sandy, a roller coaster still sits in the ocean.

Seaside Heights, New Jersey
February 25, 2013

A second-order consequence of the oceans heating up is the disruption of the global water cycle.

A storm brews above the
Atlantic Ocean.

Cape San Blas, Florida
August 2, 2012

When ocean temperatures go up, the water vapor rising from the ocean increases significantly. Moreover, warmer air can hold a lot more water vapor. For each additional 1°C (1.8°F) of temperature, the atmosphere's capacity to hold water vapor increases by 7 percent. There is already 4 percent more water vapor over the oceans than there was just 30 years ago.

THE HYDROLOGICAL CYCLE

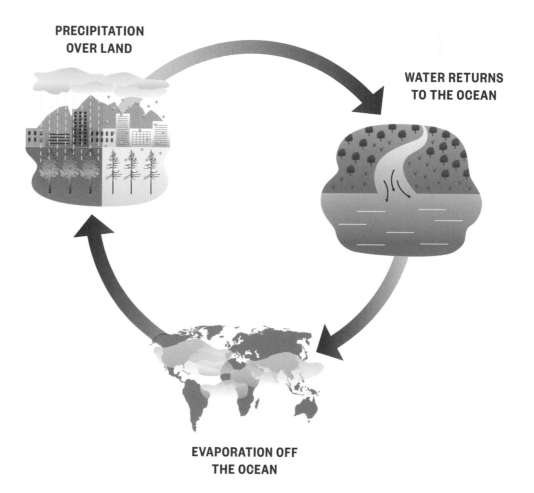

PRECIPITATION
OVER LAND

WATER RETURNS
TO THE OCEAN

EVAPORATION OFF
THE OCEAN

ATMOSPHERIC RIVER

HAWAII

This added water vapor can strengthen what are called atmospheric rivers (or "flying rivers")—flows of water vapor hundreds of miles wide that can carry 15 times as much water as the Mississippi River.

When atmospheric rivers reach the continents, they often release record-breaking downpours.

CALIFORNIA

When a complex system has many consequences and you change the system, *all* of the consequences change. That means every storm is different now because it takes place in a warmer and wetter world.

An atmospheric river drenching the western United States—seen from the NOAA/NASA Suomi NPP satellite.

—

Pacific Ocean
February 20, 2017

Water vapor is often funneled thousands of miles from the oceans over land. Then, much more of it falls all at once.

They are calling these events rain bombs.

An intense downpour drenches the American southwest.

Phoenix, Arizona
July 18, 2016

Extreme weather events are happening with much greater frequency. In a single year, Houston, Texas, was hit by two 1-in-500-year floods and a 1-in-1,000-year downpour. What used to be a 500-year flood is not that anymore.

Nayelo Cervantes helps her friend's daughter, Sophia Aviles, through the floodwaters after a huge downpour.

Houston, Texas
May 26, 2015

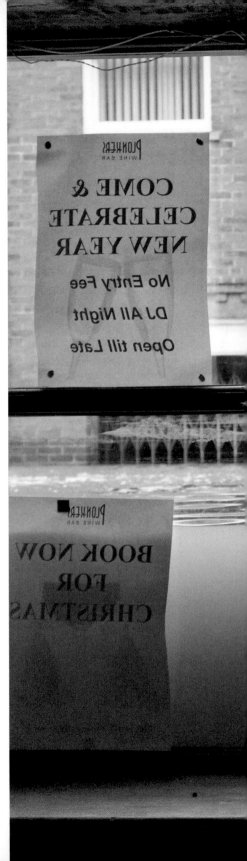

A woman cleans the inside windows of her bar after the Foss and Ouse Rivers burst their banks, causing the city to flood.

York, United Kingdom
December 28, 2015

Catherine Flowers

Founder of the Alabama
Center for Rural Enterprise

MONTGOMERY, ALABAMA

CATHERINE FLOWERS WAS a young girl in Birmingham during the Civil Rights Movement, and she vividly remembers when terrorists bombed the 16th Street Baptist Church. Her family later moved to nearby Lowndes County, in Alabama's Black Belt, where they embraced the tight-knit rural community.

"My parents had their own vegetable garden," Flowers said. "When you went to their home, it was customary to give you something from their garden."

Flowers soon noticed something unsettling about the local cotton industry, which dated back to the days of slave plantations. When the farmers sprayed their crops with DDT, she would soon see dead birds and snakes in the road nearby.

The images of these dead animals stayed with her, even after she left Lowndes County to serve in the Air Force. Years later, when she was working as a social studies teacher, she came across a map predicting how the United States would be reshaped if the polar ice caps melted. Eventually, she started to take note of devastating storms, invasive species, and summers that seemed to go on and on. She became a committed environmentalist.

"As a grandparent, I'm concerned," said Flowers, who has come to see deep connections between race, poverty, spirituality, and the environment. "That's one part of why I'm in this movement. I don't want my grandchild to wake up and realize the apocalypse was man-made."

In 2000, when she returned to Lowndes after decades away, she was horrified to discover how little the area's legacy of inequality had changed. Poverty was widespread, public health was poor, and the infrastructure was crumbling. Lowndes was particularly troubled by substandard wastewater sanitization treatment resources, which she came to suspect were related to outbreaks of hookworm and other parasitic diseases.

Flowers started going door-to-door to survey the community about sewage and sanitation. What she found shocked her: raw sewage that bubbled up into residents' bathtubs and yards when it rained, poor access to the municipal sewer, and a sewage sprayfield that overflowed into a local creek. These findings were bolstered by a 2011 United Nations report about sanitation and water quality in the United Stations that highlighted dire conditions in the Black Belt. She also noticed that the heavier rainstorms caused by the climate crisis were making the sewage overflow problems much worse.

"We found that there was no wastewater infrastructure," Flowers said. "We also found that people were being arrested because they could not afford on-site septic tanks."

Flowers founded a nonprofit, the Alabama Center for Rural Enterprise, which advocates for better

Poor wastewater infrastructure causes sewage to overflow into a creek and contaminate the local water supply.

Uniontown, Alabama
May 21, 2015

basic infrastructure in the area. In 2012, she read an op-ed in the *New York Times* by Peter Hotez, a tropical disease researcher at Baylor College, who decried domestic outbreaks of dengue fever, cysticercosis, toxocariasis, and other illnesses specific to the developing world. The two are now working together to test for tropical disease in Alabama.

"One of the most extreme examples of inequality in this country is finding tropical parasites in one of the wealthiest nations in the world," Flowers explained. "Wherever you see this in the world, you find poverty."

Flowers has also had a long-standing interest in the ways that spirituality can inform discussions of racial justice, poverty, and environmentalism. Since 2015, she's augmented her community work in Alabama with a position at Union Theological Seminary's Center for Earth Ethics, where she works with my daughter, Karenna Gore, to encourage religious communities to engage with climate issues and to create partnerships between secular and faith-based environmental leaders.

"The Center for Earth Ethics is about helping religious and spiritual leaders understand and articulate to their congregations the science of climate change," Flowers said. "I think it's been very powerful."

Last year, Flowers visited the camp at Standing Rock, where people were demonstrating against the Dakota Access Pipeline. She observed a series of ceremonies honoring water and the Earth that made a lasting impact on her. She recalled how struck she was by activists from such diverse backgrounds all coming together to celebrate the universal importance of clean and safe water. The Standing Rock Sioux captured this powerful truth with their saying "Water Is Life."

"I'm very hopeful," she said. "I'm very hopeful because we have people talking about these things now. It has brought us together in a way that's never happened before." ◉

Scientists say that the United States could see a 400 percent increase in extreme downpours by the end of this century. The same pattern is true globally.

Predictions of our future can no longer be based on our past.

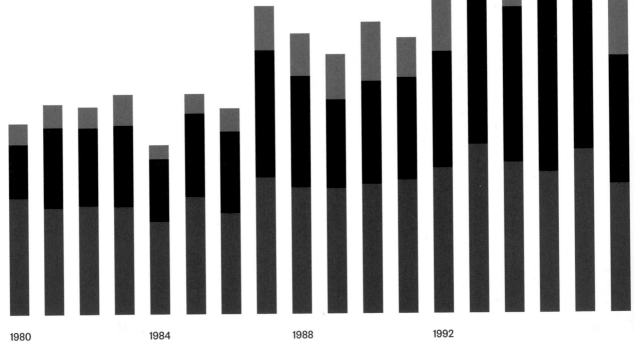

1980   1984   1988   1992

SOURCE: INSURANCE INFORMATION INSTITUTE

TRUTH TO POWER

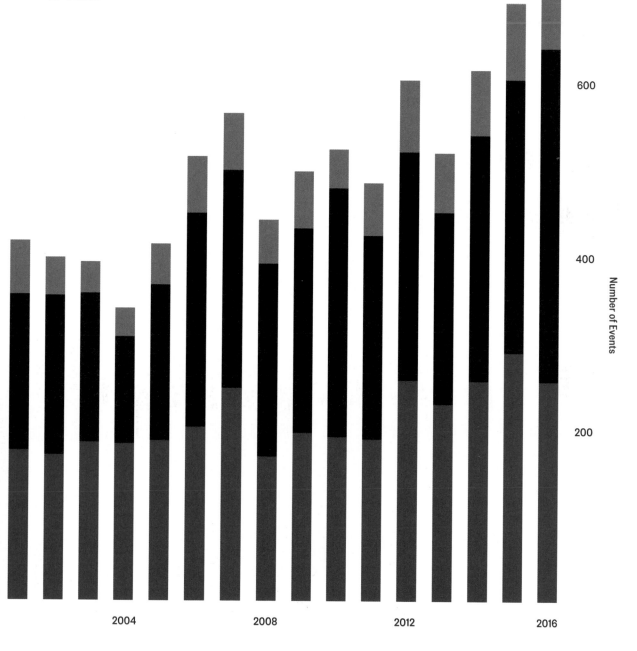

WORLDWIDE EXTREME WEATHER CATASTROPHES, 1980–2016

- Extreme temperatures, droughts, fires
- Floods, mudslides
- Storms

800

600

400

200

Number of Events

2004   2008   2012   2016

The same extra heat that is evaporating water off the oceans is also sucking moisture out of the soil, causing more droughts, deeper droughts, and longer droughts.

Livestock are often the first affected by drought due to lack of adequate grazing land.

Maras, Ethiopia
February 4, 2016

Historic droughts in eastern and southern Africa are contributing to a region-wide famine, putting 20 million people at risk. The United Nations has called this emerging tragedy the worst humanitarian disaster since 1945.

More than 400 Indian farmers committed suicide in the first four months of 2016, due mainly to pressures from the region's ongoing drought.

The dry bed of the Manjara Dam reservoir, which is supposed to supply water to several towns and villages.

—

Near Kaij, India
May 10, 2016

Where there's drought, the vegetation dries out and fires increase. Fires are becoming much larger and occurring much more frequently.

It is unprecedented.

Aggravated by drought and high temperatures, wildfires burn in the Angeles National Forest.

Los Angeles, California
June 20, 2016

A wildfire threatens Highway 63, causing drivers to abandon their vehicles.

—

Fort McMurray, Canada
May 7, 2016

In Fort McMurray, the heart of Canadian tar sands production, 100,000 people had to be evacuated from their homes because of a 1.5 million acre wildfire in 2016. The temperature in the area was 22.2°C (40°F) above normal, and most of the trees had been killed by bark beetles. All production from the tar sands came to a halt.

The amount of forestland burned in Canada has doubled since the 1970s. It's expected to double again, or even quadruple, because of climate change.

The linkage between high temperatures, drought, and fire is extremely well established.

NUMBER OF LARGE FIRES IN THE WESTERN U.S.

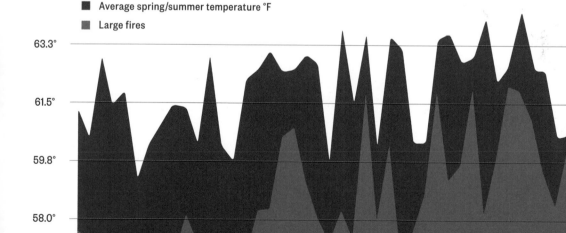

SOURCE: NOAA, US FORESTRY SERVICE, CLIMATE CENTRAL

Following a historic drought in 2010, wildfires and a record heatwave killed 55,000 people in Russia. After these fires, Russia took all of its grain off world markets and Ukraine restricted its grain exports.

World food prices hit an all-time record level for the second time in three years. These two price spikes caused food riots or civil unrest in 60 countries.

FOOD PRICE INDEX

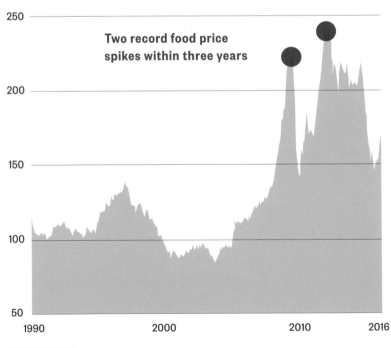

Two record food price
spikes within three years

250

200

150

100

50

1990   2000   2010   2016

SOURCE: UN FAO

A woman flees one of a wave of forest fires that devastated central Russia.

———

Near Vyksa, Russia
July 29, 2010

As food prices hit a record peak, a food vendor in Tunisia set himself on fire and touched off the Arab Spring, which began an upheaval throughout the region.

Rioters clash with police.

Tunis, Tunisia
January 14, 2011

Look at what has happened in Syria. From 2006 to 2010, the country had a record-breaking drought. Sixty percent of all the farms in Syria were destroyed. Eighty percent of all the live-stock were killed. The drought drove 1.5 million people into Syria's already crowded cities.

Syrian officials said that the effects of the drought were "beyond our capacity as a country to deal with."

The gates of hell opened in Syria. There are multiple causes for the civil war there, but the principal underlying cause was the climate-related drought (the worst in 900 years) that devastated that country.

Syrian farmer Ahmed Abdullah looks out onto a landscape that is turning barren due to drought.

—

Al Raqqa, Syria
September 23, 2010

"I had 400 acres of wheat, and now it's all desert."

—Ahmed Abdullah, Syrian farmer

The climate-related drought throughout the Middle East and North African region is contributing to a tremendous flow of refugees.

Migrants try to board a train traveling west from the Croatian border with Serbia.

Tovarnik, Croatia
September 20, 2015

"Climate change will likely lead to food and water shortages, pandemic disease, disputes over refugees and resources, and destruction by natural disasters in regions across the globe."

—U.S. Department of Defense, 2014

Migrants off the coast of Libya, hoping to be rescued by members of the NGO Proactiva Open Arms.

Mediterranean Sea
October 4, 2016

Nana Firman

Climate Reality Leader

RIVERSIDE, CALIFORNIA

WHEN NANA FIRMAN AND her husband moved to California in 2012, they attended the Islamic Center of San Diego, California, a local mosque. But in the evenings during Ramadan, Firman noticed that the congregation was using disposable dishware and throwing out a lot of food.

It bothered Firman, who has been a committed environmentalist since her days as a city planner helping rebuild her native Indonesia after the tsunami that devastated the Aceh region in 2004.

So she arranged a meeting with the mosque leadership to discuss how they could reduce waste and take climate change seriously. But since she was new to the community, she decided to prepare by poring over the Quran for verses advocating a sustainable lifestyle.

"I'm a Muslim, so I went back to my own faith to study what Islamic teachings said about protecting the environment," she said. "I used that language as a narrative."

The pitch was successful: the board of directors committed to celebrating a "Green Ramadan" the following year, and eventually ended up installing solar panels on the mosque's roof. The Imam was so impressed that he suggested Firman hold a talk about sustainability for the congregation.

The talk went over well, and Firman ended up holding similar presentations about the connections between her faith and environmentalism at five other mosques in the area. Picking up momentum, she started writing about the topic for Muslim publications and delivered a TEDx talk in France.

She also became a fellow at GreenFaith, an interfaith nonprofit that advocates for environmental leadership in religious communities.

"In a lot of ways, we have the same problems," she said of her interfaith colleagues at GreenFaith. "Everybody wants to have more efficiency in their house of worship."

The thought she has put into the connections between Islam and conservation, Firman says, has deepened both her commitment to climate progress and her understanding of her own religion.

"The prophet taught people to plant trees," she said. "If tomorrow is the end of the world, and you have a seedling in your hand, you should plant it. You should always have hope." ◉

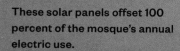

These solar panels offset 100 percent of the mosque's annual electric use.

San Diego, California
October 2016

When any great moral challenge is ultimately resolved into a binary choice between what is right and what is wrong, the outcome is foreordained because of who we are as human beings.

When we upset the balance of nature, we must understand that the consequences can be more severe than simply higher temperatures. Many interrelationships are profoundly affected.

An example is our globally integrated food system. The crops we eat today were patiently selected over hundreds of generations during the Stone Age. These food crops thrive in the natural conditions in which they evolved. Now that we are changing those conditions, many of these crops are becoming stressed, especially by higher temperatures. And they are not giving us the same yields or nutrient quality.

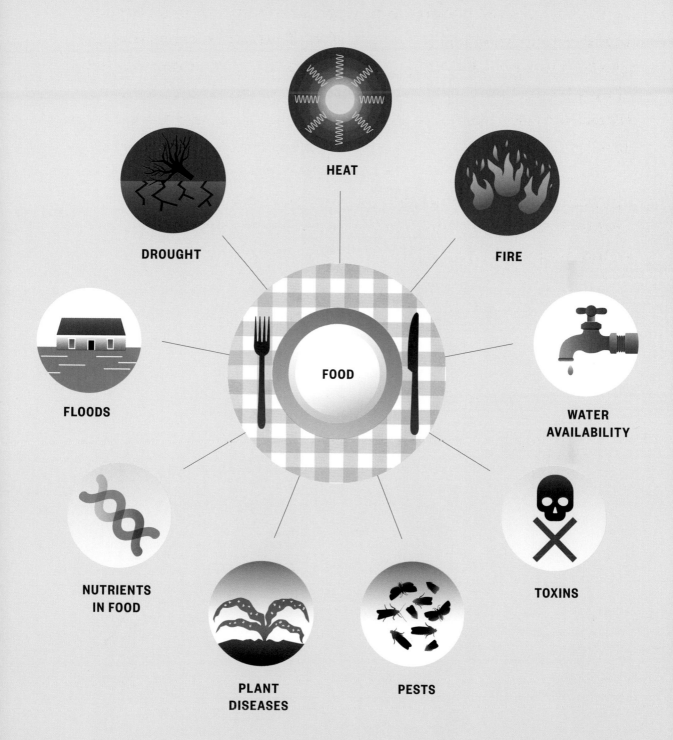

Research shows that when temperatures go above a certain level, it causes significant drops in food crop yields. In China, we have already seen a decline in wheat and corn yields of 5 percent in the past three decades. In the U.S., the yields for corn could drop by one-third from heat stress alone. Wheat yields could drop by more than one-fifth.

A Mexican farmer shows corn that has died due to lack of rain.

—

Pocoboch, Yucatan

GLOBAL WATER USE

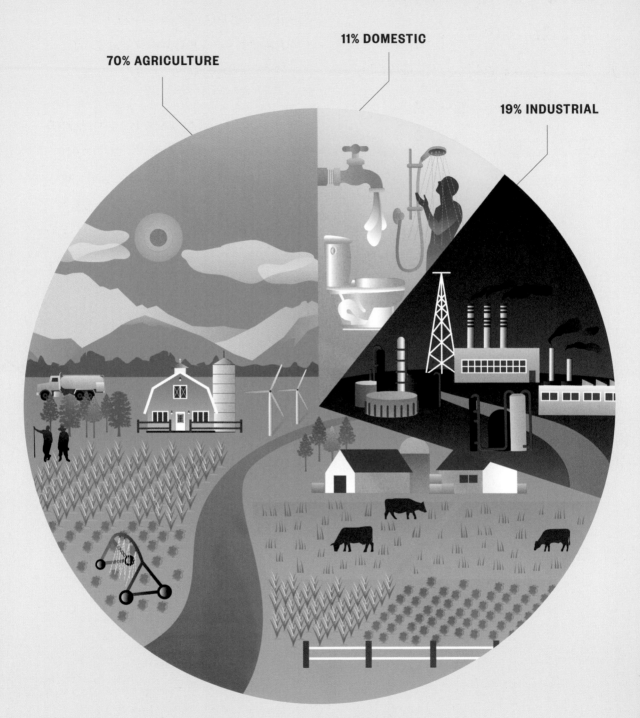

70% AGRICULTURE

11% DOMESTIC

19% INDUSTRIAL

SOURCE: AQUASTAT

Everything—people, crops, energy production, industry, and animals—requires more water when the temperature goes up.

In many areas of the world, water scarcity is one of the single most serious aspects of the climate crisis.

Higher temperatures increase the demand for water. About 11 percent of consumption of the world's freshwater is used in our homes, about 19 percent by industry, and 70 percent by agriculture.

Cows roam a drought-ridden landscape at a cattle range in California's Central Valley.
—
Delano, California
February 3, 2014

The even more complex systems of human and public health are also being stressed by climate change, which has been declared a global health emergency by the prestigious medical journal *The Lancet*. We have recently seen some of the dangers scientists have predicted beginning to get more severe.

"The effects of climate change are being felt today, and future projections represent an unacceptably high and potentially catastrophic risk to human health."

—*The Lancet Commission on Health and Climate Change*, April 2015

THE IMPACT OF THE CLIMATE CRISIS ON HUMAN HEALTH

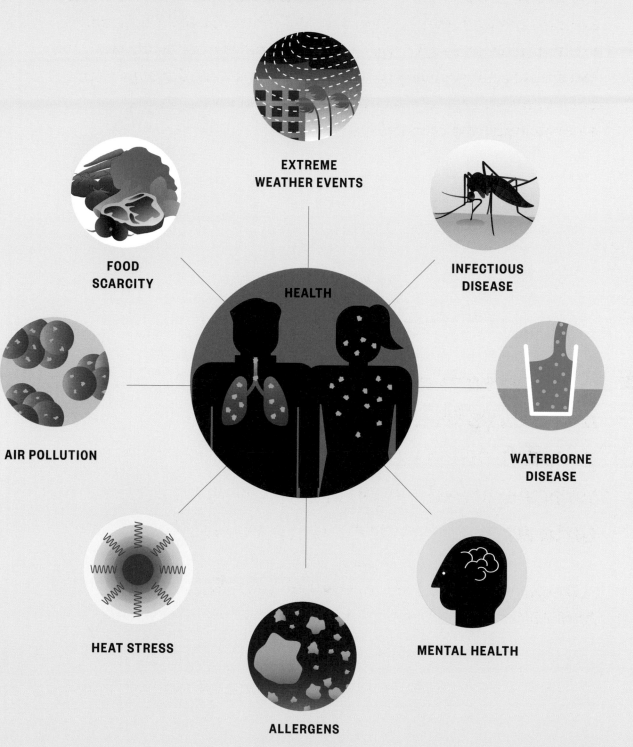

EXTREME
WEATHER EVENTS

FOOD
SCARCITY

INFECTIOUS
DISEASE

HEALTH

AIR POLLUTION

WATERBORNE
DISEASE

HEAT STRESS

ALLERGENS

MENTAL HEALTH

Susan Pacheco

Pediatrician and Climate Reality Leader

HOUSTON, TEXAS

DRAWING ON HER DECADES of experience as a pediatrician, Susan Pacheco has a simple message for doctors everywhere: the health of the environment affects the health of their patients.

She points to research associating heat waves with elevated numbers of emergency room visits, air pollution with developmental problems in children, and natural disasters with population-wide spikes in depression, anxiety, and post-traumatic stress disorder. And those ailments, she worries, could be just the tip of the iceberg.

"There's heart disease, there's lung disease, there's kidney disease," said Pacheco, a professor of pediatrics at the University of Texas McGovern Medical School. "Every organ system can be affected by climate change. When I say that, I get goosebumps."

Pacheco, who was born and raised in Puerto Rico, didn't become concerned with climate science until 2006. Her eldest son was learning about climate change in school, so she took the family to see *An Inconvenient Truth*.

This trip to the theater proved to be a wake-up call. She had never paid much attention to climate science, but after seeing the movie she found herself preoccupied by it. As time passed, she decided she needed to take action, and applied to take part in the second-ever Climate Reality Leadership Corps, a training program I led in Nashville in 2006.

It wasn't until she returned home to Houston and started giving her own presentations that Pacheco found her niche. She began to familiarize herself with the body of research on how the climate affects health, and she was startled by the gravity of what she learned.

Beyond the obvious concerns about extreme heat and pollution, researchers worry that extended allergy seasons could disrupt the functioning of our immune systems, and higher temperatures allow mosquitoes and ticks to spread diseases for a longer period than ever before. High carbon dioxide levels can even affect the nutritional value of crops.

Pacheco became convinced she could see the effects in her own clinic's waiting room, in the Texas children she saw suffering from asthma, heat sensitivity, and allergies. Children and the elderly, she discovered, tend to be the most vulnerable. And while many adults have lived for years in an environment less affected by climate change, today's youth will grow up with an entire lifetime of exposure. The potential for damage and illness, she suspects, is much higher.

She decided that those vulnerable populations needed an advocate inside the health care system, and she started to adapt her presentation for the task of educating other physicians about the health implications of the changing climate. "The

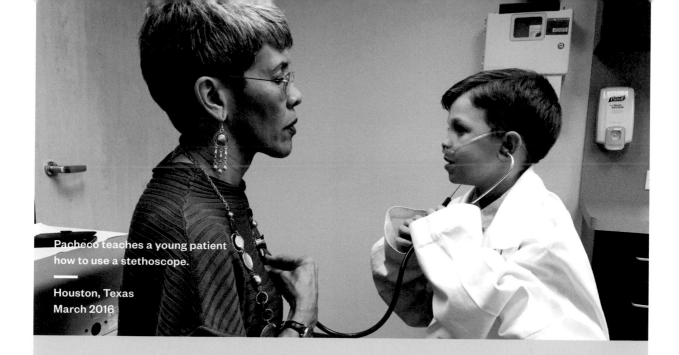

Pacheco teaches a young patient how to use a stethoscope.

Houston, Texas
March 2016

knowledge about climate change in the health community was almost nonexistent," she said. "So that's what I've been doing since that time: working to educate and build awareness in the health sector."

In the years since, she has spoken in dozens of professional settings, to ethnic and religious organizations, to family groups, and, above all, to health care practitioners. At first, some fellow doctors resisted her message. Her own father suggested, pointedly, that she spend more time focusing on her medical career and less on the environment.

But as time went on and the national conversation around the climate shifted, Pacheco found the medical community to be more and more receptive. A key victory has been that medical schools have started to integrate the health effects of climate change into their curricula, starting with her own students at McGovern.

Pacheco also founded the Texas Coalition for Climate Change Awareness, a network dedicated to educating communities across the state about the threats of climate change, and penned newspaper editorials in both English and Spanish about the importance of EPA regulations. She worked with the American Academy of Pediatrics' Council on Environmental Health to publish a policy statement about the importance of considering the environment in pediatric health.

Some authorities have started to take notice of Pacheco's efforts. Last year, 13 federal agencies collaborated on a report concluding that climate change is a "significant threat" to the health of Americans.

"The changes are happening right now," said former U.S. surgeon general Vivek Murthy. "Climate change is going to impact health, and it's not a pretty picture."

In 2013, the White House bestowed Pacheco with the illustrious "Champions of Change" award in recognition of her efforts. "I want to educate," she said. "I will talk to as many people as I can." And she doesn't intend to rest until the entire medical establishment has taken note. ◉

See page 302 for advice on adapting the Climate Reality presentation for your own passions.

One major consequence of the climate crisis is that the balance between human beings and microbes is being upended. Climate change means the "vectors" that carry diseases, including mosquitoes and ticks, have a wider range. In central China, the reemergence of malaria is directly related to increased temperatures and disruptive rainfalls.

In the United States, the range of the *Aedes aegypti* mosquito is extending northward. This is the principal mosquito that carries the Zika virus.

In addition, the life cycles of both *Aedes aegypti* and the Zika virus have been accelerated by warmer temperatures, increasing the time available for transmission of the virus.

AEDES AEGYPTI LIFE CYCLE

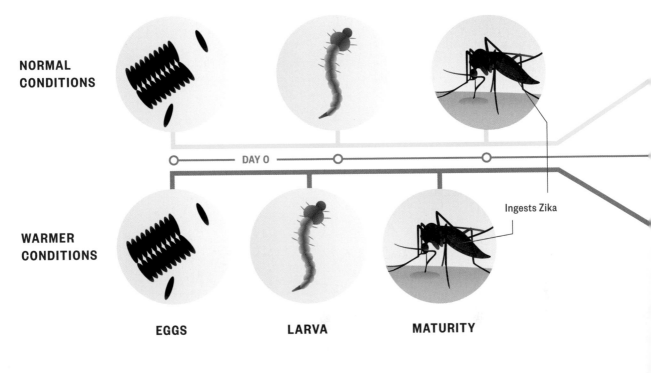

NORMAL CONDITIONS

DAY 0

WARMER CONDITIONS

Ingests Zika

EGGS   LARVA   MATURITY

CURRENT RANGE OF *AEDES AEGYPTI* IN THE CONTINENTAL UNITED STATES

■ *Aedes aegypti* range

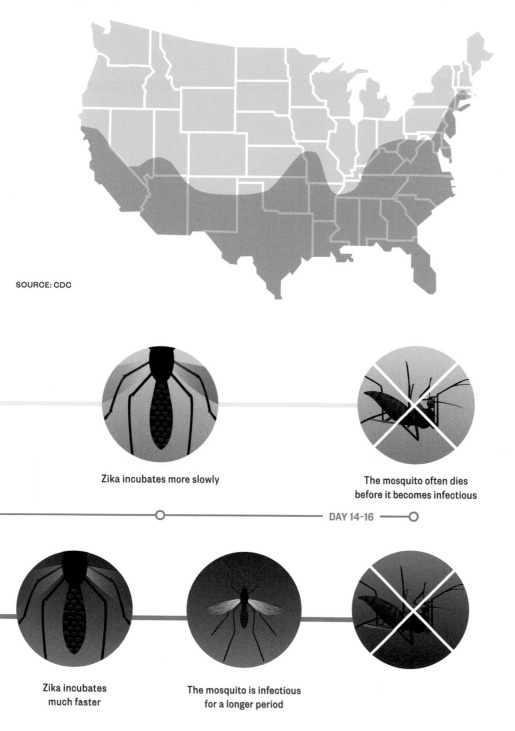

SOURCE: CDC

Zika incubates more slowly

The mosquito often dies
before it becomes infectious

DAY 14-16

Zika incubates
much faster

The mosquito is infectious
for a longer period

In several countries in South America and Central America affected by Zika, doctors advised women not to get pregnant for two years. Telling women that they should not get pregnant until we have this under control—that is a message that is new in the history of the human race.

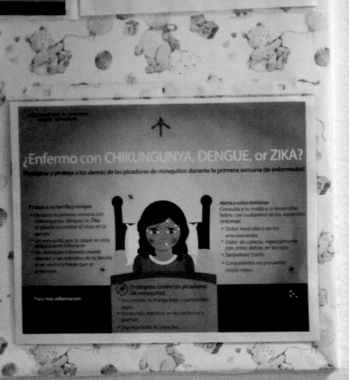

¿Enfermo con CHIKUNGUNYA, DENGUE, or ZIKA?

At 37 weeks pregnant, this mother is at risk of her baby having birth defects due to the Zika virus.

—

San Juan, Puerto Rico
February 3, 2016

The burning of carbon-based fuels creates not only greenhouse gases but also conventional air pollution. India and China are among the nations facing great health challenges from this dirty air.

Worldwide, air pollution kills 6.5 million people every year.

Electricity towers on a smoggy day.

—

New Delhi, India
November 30, 2015

Ivy Chipasha

Climate Reality Leader

LUSAKA, ZAMBIA

IVY CHIPASHA WAS BORN in Zambia's Copperbelt region, where mineral and emerald mines dot the landscape. She later moved to the country's Central Province, where swamplands surround the mouths of the Lukanga and Kafue Rivers, and finally to the relatively cosmopolitan Lusaka area, where she raises four children. In each place, she has witnessed climate change ravage her homeland.

She has seen deforestation menace the country's woodlands, which are among the most prolific in Africa. Droughts now alternate unpredictably with heavy rainfall and flash floods, and the irregular weather has wreaked havoc on the nation's hydropower infrastructure, causing outages and power shedding.

"We've been hit hard by climate change," she said ruefully.

Chipasha originally came to environmentalism by way of her work in the financial sector, where she became convinced that Africa's energy future lay in renewables. In 2014, she attended the Climate Reality Leadership training I led in Johannesburg, South Africa, and in the years since, she has given many presentations, often to students in the Lusaka area.

She is also a steering committee member of the African Renewable Energy Alliance, an organization that helps policy makers, business leaders, and academics make plans for the continent's energy future.

"If leeway is given to climate change activists, a lot will change," she said. "Climate change affects everyone."

Because the Zambian government has limited financial resources, Chipasha is also interested in ways that education and outreach can help individual families conserve. In rural parts of Zambia, families often still cook using firewood and charcoal, and the growing demand is putting a strain on the country's lush forests.

To promote renewable energy in Zambian households, she founded an organization called the Green Environment Foundation, which advocates for clean water, conservation, and lower-impact cooking technologies in Zambian households. Chipasha was a peer reviewer of "Beyond Fire," a report by the World Future Council that was presented at the United Nations Framework Convention on Climate Change in Morocco last year.

One day, Chipasha believes her work will lead to a brighter environmental future for her children and her country.

"I believe that we will win this fight," she said. "Nothing is impossible." ◉

A baby with respiratory disease
undergoes inhalation therapy.

Beijing, China
January 28, 2013

The pollution in 80 percent of China's cities exceeds what's considered safe by air-quality standards. In Northern China, life expectancy has gone down 5.5 years because of air pollution.

"At the present time ... Beijing is not a livable city."

—Wang Anshun, former mayor of Beijing, January 23, 2015

Pollution Monitoring

IN JULY 2008, THE U.S. EMBASSY in Beijing connected its rooftop pollutant monitor to a new Twitter account and began automatically tweeting its data. Using the Twitter handle @BeijingAir, the hourly tweets followed a simple and uniform format: date, time, pollutant type, concentration, and air quality index (AQI).

Beijing's air quality was already notoriously poor. However, despite residents' frequent inability to see the full city skyline due to smog, official government pronouncements listed the city's air pollution as "light." The discrepancy between official government pronouncements and Beijing residents' day-to-day reality was due to the type of air particles the government monitored.

The most dangerous type of air pollutants are PM2.5—tiny particles that measure less than 2.5 microns in diameter, or about the width of spider silk. These particles are small enough to enter the lungs and even the blood, where they can wreak havoc and potentially cause serious heart and lung problems. The Chinese government monitored PM10 pollutants, which are much larger and much less dangerous. Because of the choice of particle size, Beijing's obvious air pollution problem was not reflected in the official measurements by the Chinese government.

In November 2010, the U.S. Embassy sent out a tweet describing Beijing's air quality as "crazy bad."

The AQI measures particle pollution on a scale of 1–500, with 301–500 being "hazardous." That day, the score was quite literally off the charts, at 562. The tweet went viral, and although Twitter is blocked for most users in China, screenshots of it circulated on other platforms, sparking an uproar in Beijing.

The brilliance of the tweets was their simplicity. The account was only tweeting out data. Aside from the "crazy bad" categorization, @BeijingAir was not tweeting analysis or declarations. The account left its tweets to be interpreted by the reader, therefore it was able to avoid being declared subversive. The embassy was allowed to continue its pollution monitoring and tweeting, notwithstanding the fact that their data directly contradicted the official government story.

Local environmental advocates seized upon this new information. Using the momentum of the viral tweet, they mobilized Beijing residents to monitor air quality themselves and to advocate for changes. The Green Beagle Environment Institute began distributing PM2.5 monitors to local residents. These monitors were about the size of a transistor radio, and much less expensive than the embassy's, but they produced the same data.

By 2013, environmental advocates had succeeded in convincing China to set up hundreds of PM2.5 monitoring stations in over 70 cities.

U.S. Embassy air quality tweet.
Twitter, January 12, 2013

BeijingAir @BeijingAir
01-12-2013 17:00; PM2.5; 810.0; 704; Beyond Index
Expand

Additionally, China earmarked hundreds of billions of dollars for air cleaning and began implementing policies for air pollution reduction targets in major cities, most importantly adopting the PM2.5 measurement for data.

Today, Beijing's AQI rating hovers around the "unhealthy" score, 151–200. While there is clearly more work to be done, it's a huge change from the "hazardous" and off-the-charts ratings.

Additionally, the increased access to daily data from the monitoring stations helps residents make important decisions about everyday activities, like how long their kids can play outside, or if it's safe to let elders go out for walks. In an age when any news source can be branded "fake," providing the public with raw, objective data can be a subversive act. ◉

The Dongbianmen Gate before and after the smog sets in. Beijing, China, January 12, 2013 (before) and January 16, 2013 (after).

Wang Shi

Founder and Chairman of China Vanke

SHENZHEN, CHINA

WHEN WANG SHI WAS YOUNG, he was drawn to stories of adventure, particularly Jack London's *The Call of the Wild* and Ernest Hemingway's *The Snows of Kilimanjaro.*

And it was on Kilimanjaro, years after he became one of the most successful real estate entrepreneurs in China, that Wang experienced a wake-up call about climate change. In 2002, when his climbing party reached the peak, he was struck by the lack of snow. When he returned home and researched the topic, he discovered that in 20 years the mountain's snows could be gone entirely.

"I began to notice more signs of climate change, as I carried on my adventure journeys in other parts of the world," said Wang, who, in addition to climbing the highest peaks on every continent, also serves as the chairman of Vanke, the largest residential real estate developer in China.

Wang began reading more about climate science. He found that China produces the most carbon emissions of any country in the world, and that the nation's construction industry, where he'd built his career, was responsible for more carbon emissions than Russia and India combined.

After a time, Wang came to believe that he bore a responsibility, as a prosperous businessperson, to advocate for environmentalism at home and abroad. "Although China is growing quickly into the second largest economy, the industrialization process is taking on big environmental and social prices," he said. "I think the current growth model cannot continue in the long term. Chinese companies and entrepreneurs are at a crossroads and need enlightenment for a more sustainable and healthy model."

The first step was to take strides to make Vanke's work more sustainable. Wang was drawn to prefabrication, a construction technique that assembles as many components of a structure as possible in a centralized manufacturing location. While often more costly, prefabrication uses far less timber, water, and energy than conventional methods.

At first, Wang had little support within either the industry or the government. Some at Vanke were unconvinced. But as it became clear that the tide in the construction sector was turning toward prefabrication, Vanke found itself in an auspicious position.

"When the government later announced standards and incentives for implementing new techniques, we were already taking the lead in the industry and did not have difficulty in adjustments," he said. "Our early move was rewarding."

Wang went on to found a group of Chinese entrepreneurs called the C Team, with whom he is working to network business leaders in the country to fight climate change. The group has attended the United Nations' climate conferences almost

TRUTH TO POWER

The prefabrication lab at Vanke's national R&D center.

Dongguan, China

every year since, and last year brought more than 100 Chinese business leaders to the summit in Marrakech, Morocco.

Wang has also advocated for the health of mountains. In 2010, in conjunction with his second trip to the summit of Mount Everest, the Vanke Foundation organized a campaign called Kilometer Zero that called on Chinese mountaineers to bring trash down from mountainsides. The effort collected some two tons of garbage, including 200 abandoned oxygen tanks.

In recent years, much of Wang's work has been in the realm of higher education. He has taught at the Hong Kong University of Science and Technology and Peking University, and in 2011 he joined the Harvard University Asia Center to study connections between business ethics and Judaism. While at Harvard, he visited Walden Pond and read the works of Henry David Thoreau to reflect on the importance of the natural world.

"It is not the question of *how*, but the question of *why* we should ask ourselves," Wang explained. "The purpose of my study is to bring back inspirations to China from the foundation of Western industrialization."

Last year, Wang took part in the Climate Reality Project's second-ever training in China. He spoke about the ways climate change could impact China's nearly 1.4 billion citizens and the ways that the country can act as a global leader for climate progress.

"I am a cautious optimist," Wang said, not just for the future of the mountaintops he loves, but also because of the growing commitment to environmentalism he sees in China's public and business spheres. ◉

The balance of nature is also visibly upset in the Arctic. Here, during the long polar winter night, the sun doesn't hit the North Pole for six months of the year. Normally, it's extremely cold.

But these are not normal times. Right in the middle of that polar winter night, on February 10, 2017, temperatures were about 27.7°C (50°F) above normal. The North Pole started thawing at night in the middle of winter. A similar temperature spike occurred in December 2015. This is a pattern.

TEMPERATURE DEPARTURE FROM AVERAGE
FEBRUARY 10, 2017

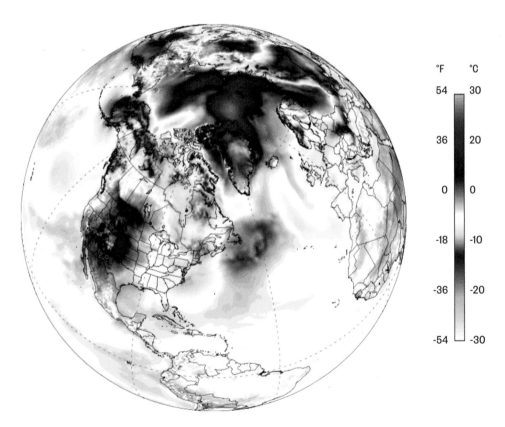

SOURCE: CCI, UNIVERSITY OF MAINE

The Arctic is already undergoing dramatic changes.

As underwater ice melts, methane
trapped beneath bubbles up and
is released into the atmosphere.

———

Cook Inlet, Alaska
January 2010

One of a number of methane craters in Siberia.

Yamal Peninsula, Siberia
August 25, 2014

This is not where a meteor has hit.

It is an outburst crater in Siberia caused by a methane explosion after the tundra thawed and gas built up until it blew out this massive hole.

Throughout the Arctic tundra, huge amounts of plant matter lie frozen in the soil. When it thaws, the decomposition produces some CO_2, and more worryingly, enormous amounts of methane, which is a far more potent greenhouse gas.

We don't yet know how dangerous the release of methane in the warming Arctic will be, or when we might cross a tipping point when there will be a much larger release of methane.

Konrad Steffen

Climate Scientist and Climate Reality Leader

EGG, SWITZERLAND

IN THE YEAR 2000, when my friend Koni Steffen looked at the latest temperature readings from the system of weather stations he'd built in Greenland, he thought something must have gone wrong with the sensors. The temperatures seemed far too high. As a good scientist, he immediately set out to corroborate the new readings.

"I went through it all night without sleeping," he recalled. "And I realized that the data was real."

Further confirmation came later after one of the stations collapsed when the ice supporting it melted rapidly in the Arctic's rising temperatures.

Steffen, whom I've known now for nearly two decades, never set out to study climate change. He was born in Zurich, Switzerland, and originally studied electrical engineering before discovering a passion for earth science.

His most long-standing project has been to monitor the weather patterns and temperatures of the Arctic using a network of solar-powered weather stations he constructed starting in 1976. He has visited Greenland to maintain and expand the network every year since, even during a lengthy period teaching at the University of Colorado during the 1980s and 1990s. This past season marked his 40th trip.

Steffen wears a stately beard, and though his hair is greying, he speaks with the ardor of a much younger man. He sometimes quips that when he started his academic career in the 1970s, scientists knew more about conditions on nearby planets than they did about Greenland.

Though he's too modest to say it, his work has been instrumental in changing that. Looking back, he's grateful for the opportunity to track climate change in the Arctic since an era when the climate was far more stable—and to have helped inform the international conversation about global warming almost since its inception.

In the years since the weather station collapsed, Steffen's research has convinced him that climate change poses a grave risk to the citizens of the world. In particular, his work has demonstrated that even small changes to the environment can have momentous climatological effects. He has shown that Greenland's ice sheet, long thought to melt more slowly than glaciers, are disappearing far more rapidly than expected, in part because of fissures that drain water down into the bedrock.

In spite of it all, Steffen has been heartened by the growing awareness about climate change over the past decade.

"We are all in it together," he said. "It is a global climate. I am a scientist, and science has no border." ◉

Antarctica Is Warming Too

MOST OF THE ICE ON PLANET EARTH is found on Antarctica. While the Arctic is ocean surrounded by land, Antarctica is land surround by ocean. In the middle of Antarctica, the surface of the ice extends 10,000 feet above sea level. When I visited the South Pole, I got altitude sickness.

The ice of Antarctica has begun to melt and flow toward the ocean at a faster pace, contributing to global sea level rise. The most vulnerable part of the ice cap is in West Antarctica, a region that is roughly the size of Greenland.

A major portion of West Antarctica has now crossed a tipping point, according to scientists, and will eventually break up, flow into the ocean, and melt. Most scientists believe that humankind still has the opportunity to prevent the breakup and melting of the bulk of ice in Antarctica, but they warn us that we must start sharply reducing global warming pollution as quickly as possible. ◉

A penguin navigates a crack in the ice. Rilser-larsen Ice Shelf, Antarctica.

The Arctic is warming faster than any other place on Earth. The melting of its glaciers and the Greenland ice cap is now raising sea levels worldwide.

BELOW: Konrad Steffen's research base on May 22, 2012.

RIGHT: The same research station 4 years later on July 11, 2016.

At a Climate Reality training in Miami, I saw fish from the ocean swimming in some of the streets in nearby Miami Beach. It was an example of "sunny day flooding." There was no rainfall, the sea was simply coming up through the storm sewers during high tides. This is happening on a regular basis.

PROJECTED SEA LEVEL RISE IN SOUTH FLORIDA

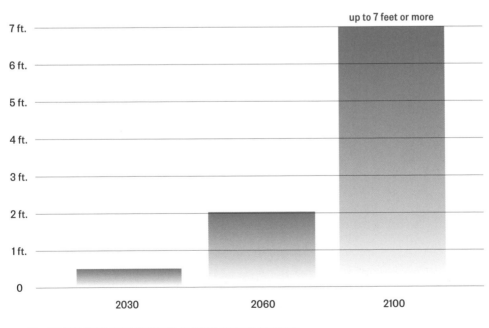

up to 7 feet or more

7 ft.	
6 ft.	
5 ft.	
4 ft.	
3 ft.	
2 ft.	
1 ft.	
0	

2030   2060   2100

SOURCE: SOUTHEAST FLORIDA REGIONAL CLIMATE CHANGE COMPACT

Miami is now the number one city in the world in terms of assets at risk due to sea level rise.

Here's something you don't see everyday: an octopus in a parking garage.

Miami Beach, Florida
November 14, 2016

Sea level rise is an extremely grave threat, both for economies and for the people who will have to relocate.

TOP 10 CITIES AT RISK FROM SEA LEVEL RISE BY 2070

BY POPULATION AT RISK

Kolkata	14 million
Mumbai	11.4 million
Dhaka	11.1 million
Guangzhou	10.3 million
Ho Chi Minh City	9.2 million
Shanghai	5.5 million
Bangkok	5.1 million
Rangoon	5 million
Miami	4.8 million
Haiphong	4.7 million

= one million people; estimated population

BY ASSETS AT RISK

Miami	$3.5 trillion
Guangzhou	$3.3 trillion
New York/Newark	$2.1 trillion
Kolkata	$2 trillion
Shanghai	$1.8 trillion
Mumbai	$1.6 trillion
Tianjin	$1.2 trillion
Tokyo	$1.2 trillion
Hong Kong	$1.2 trillion
Bangkok	$1.1 trillion

estimated assets

SOURCE: NICHOLS, ET AL., 2007, OECD

Itzel Morales

Climate Scientist and Climate Reality Leader

DAVIS, CALIFORNIA

ITZEL MORALES'S FAMILY has lived on the Mexican island of Carmen for generations. In 2010, for the first time, they had to buy air-conditioning units because a window and fan were no longer enough to keep cool during the sweltering summer months.

"The nights were suffocating," Morales said. "That fear was what got me involved in climate change."

Spurred to action, Morales enrolled in a master's program at Scotland's Heriot-Watt University, where she studied the impacts and mitigation of climate change. Then she applied to the Climate Reality Project's 2013 training in Chicago, where she participated in an intensive three-day session of climate researchers and communications experts.

After the training, Morales returned to Mexico and flung herself into climate advocacy. "I knew I

Morales and the Colectivo Ambiental Isla Verde distribute recycled notebooks to students.

could do presentations," she said. "I started sending emails to everybody I knew."

The community responded. Morales spoke in living rooms, in every elementary school in her district, and at her alma mater in Yucatán, where she had studied chemical engineering. She even presented in front of a crowd of more than a thousand at an event organized by the Fundación Pablo García, a Mexican scholarship program. In her first year, all told, she gave 25 presentations to thousands of people.

She also became more involved with local issues. Carmen Island sits on the Laguna de Términos, a tidal lagoon on the Gulf Coast where the mangroves shelter a rich ecosystem. Morales is quick to rattle off statistics about the area's abundant biodiversity—and about the looming threats of the petroleum industry. To protect the area, she became a campaign organizer for the Green Island Environmental Group, which advocates for the protection of the lagoon.

Now, she has left the region again, this time for a position as a Hubert H. Humphrey Fellow at UC Davis, where she is studying natural resource conservation and how community empowerment can push back against climate change.

She often thinks about the scorching nights back home on Carmen, but in spite of it all, she remains hopeful. The fight, she believes, is just beginning.

"We definitely have political will," she said. ◉

Our global civilization is now at a point of decision.

But there is good news.

We have all the solutions that we need right before us.

One day in August 2016, Scotland got 100 percent of its electricity from wind. In Portugal, they had four days straight in May 2016 on renewable energy alone.

The transition to renewable energy represents the largest business opportunity in the history of the world. The projections for the future are quite stunning.

The windy hills north of Lisbon are an ideal spot for a wind farm.

—

Sobral de Monte Agraço, Portugal
August 13, 2015

The best projections 15 years ago were that global wind energy would increase by 30 gigawatts by 2010.

WIND POWER PROJECTED VS. ACTUAL

Projected for 2010: 30 GW

By 2016, we beat that goal 16 times over.

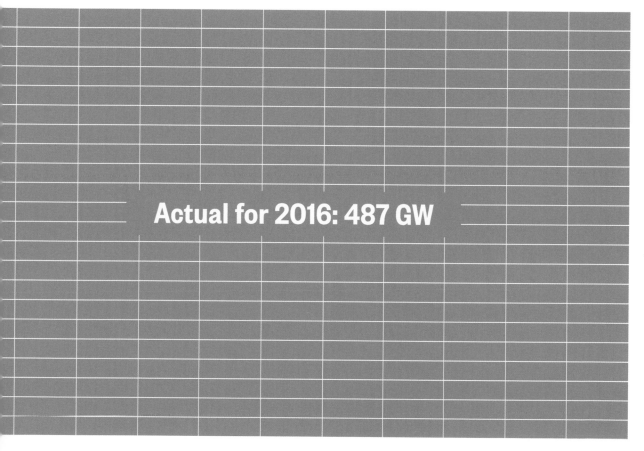

Actual for 2016: 487 GW

SOURCE: IEA, GWEC

The Middelgrunden offshore
wind farm.

Copenhagen, Denmark
March 11, 2015

For 24 hours on April 17th, 2016, wind power met 103.6 percent of Denmark's needs.

Dale Ross

Mayor

GEORGETOWN, TEXAS

JUST NORTH OF AUSTIN, Texas lies Georgetown, a city of about 50,000 known for its Victorian architecture and serene natural landscape. In many ways, it's a conventionally run municipality that tends to attract retirees—except that it's powered entirely by wind and solar energy.

More surprising, perhaps, is that Georgetown mayor Dale Ross doesn't see that incredible accomplishment as an environmental issue. To him, it's just dollars and cents.

"I think people make decisions that are in their own best interest," said Ross, who has lived in Georgetown since the fourth grade. "If you can get things so that people want sustainable green energy, I think you'll get buy-in from across the United States."

There are a handful of other U.S. cities that have made similar decisions, including Burlington, Vermont, and Aspen, Colorado, but Georgetown stands out for the overtly pragmatic language its leaders have used to justify the transition.

"There are a lot of cities that just go with the next budget cycle," Ross said, "but we plan for 25 years in advance."

We could probably all learn a lot from Ross, who spoke with an understated Texas accent as he outlined the thought process behind the transition. Ross worked for decades as an accountant, and he sees his role in government as an extension of that bookkeeping experience.

Georgetown, he explains, has run its own municipal electric utility since the early 20th century. With its wholesale power contract set to end in 2012, Ross realized the city was faced with a historic opportunity. Wind farms had been cropping up in Texas for years, and he was intrigued by the possibility of negotiating renewable rates that wouldn't rise over time.

So he crunched the numbers, examining the wind profile and solar radiance of the region, and what he found surprised him. The combination of falling photovoltaic prices and growing wind infrastructure meant that the wind and sun could not only provide all the community's electricity, but that the town could sell the extra power back to the grid. Unlike coal and oil, which are subject to price fluctuations, the city knows exactly what the costs are going to be in perpetuity.

Furthermore, he found that because fossil fuel power plants use steam, moving toward wind and solar would also reduce strain on the state's water usage during punishing droughts, while improving air quality. Sourcing the entirety of Georgetown's energy needs from renewable resources wasn't just possible, Ross realized—it also made sound financial sense. ◉

Turbines get bigger and bigger, and there are so many of them that in many areas there is often a local surplus of energy from both wind and solar. In Texas, some utilities are introducing a new rate plan. They have so much wind energy that from 9 p.m. until 6 a.m., you can use all the electricity you want totally for free.

Wind alone could supply 40 times all of the energy that the entire world needs.

A rancher cleans a water tank at the Lone Star Wind Farm, which is also a cattle ranch.

Near Abilene, Texas
June 9, 2007

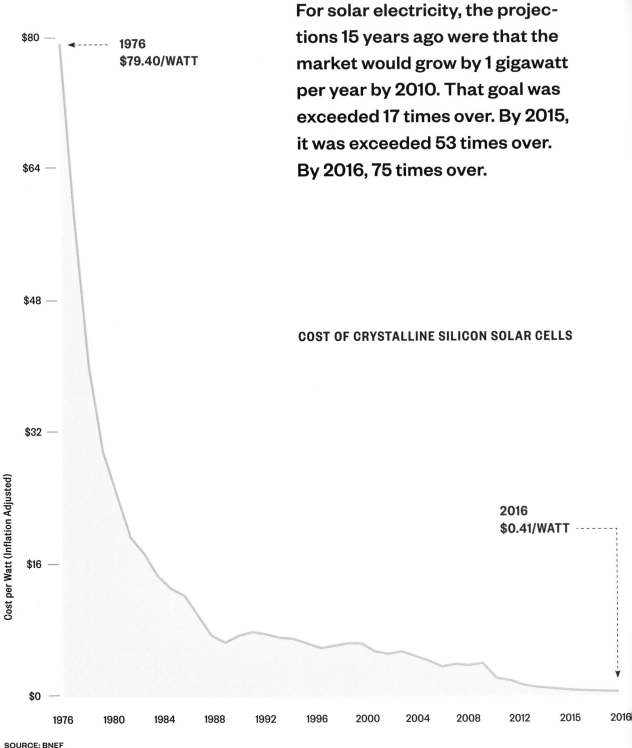

For solar electricity, the projections 15 years ago were that the market would grow by 1 gigawatt per year by 2010. That goal was exceeded 17 times over. By 2015, it was exceeded 53 times over. By 2016, 75 times over.

COST OF CRYSTALLINE SILICON SOLAR CELLS

1976
$79.40/WATT

2016
$0.41/WATT

Cost per Watt (Inflation Adjusted)

$80

$64

$48

$32

$16

$0

1976   1980   1984   1988   1992   1996   2000   2004   2008   2012   2015   2016

SOURCE: BNEF

This is an exponential curve, and it continues to go up at a steeper rate because the cost of silicon solar cells continues to go down.

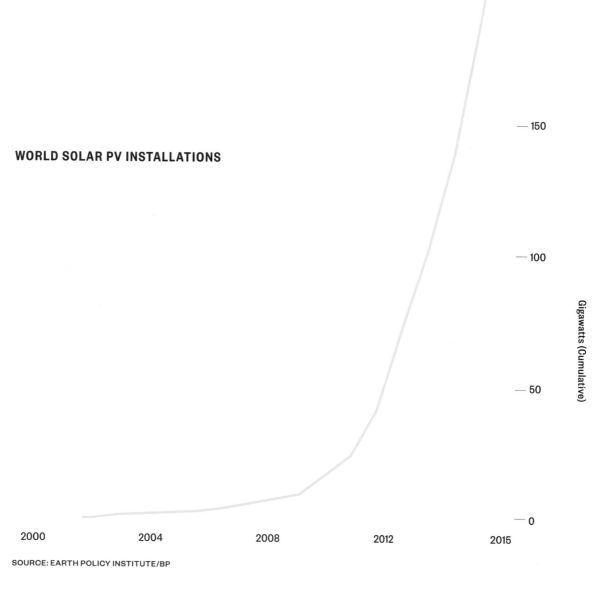

WORLD SOLAR PV INSTALLATIONS

Gigawatts (Cumulative)

250

200

150

100

50

0

2000　　　2004　　　2008　　　2012　　　2015

SOURCE: EARTH POLICY INSTITUTE/BP

Chile has great policy. It was the first South American country to enact a carbon tax.

This 100 MW solar farm is one of the largest in Latin America.

Copiapó, Chile
June 2014

At the end of 2013, Chile had 11 megawatts of solar capacity. By the end of 2014, more than 400 megawatts. By the end of 2015, more than 800 megawatts. So look at what they had under construction as of 2016, and under contract to soon begin construction. You might have to turn the page.

CHILEAN SOLAR MARKET

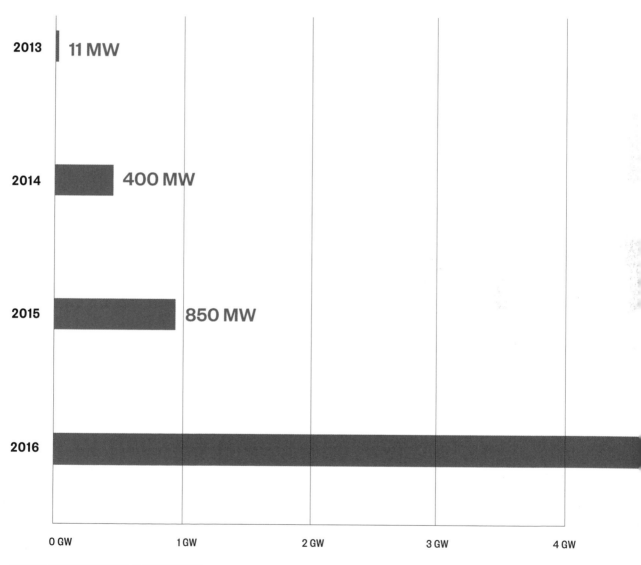

2013   11 MW

2014   400 MW

2015   850 MW

2016

0 GW   1 GW   2 GW   3 GW   4 GW

SOURCE: GREENTECH MEDIA, CLEANTECHNICA

You talk about excitement; this story gets me excited.

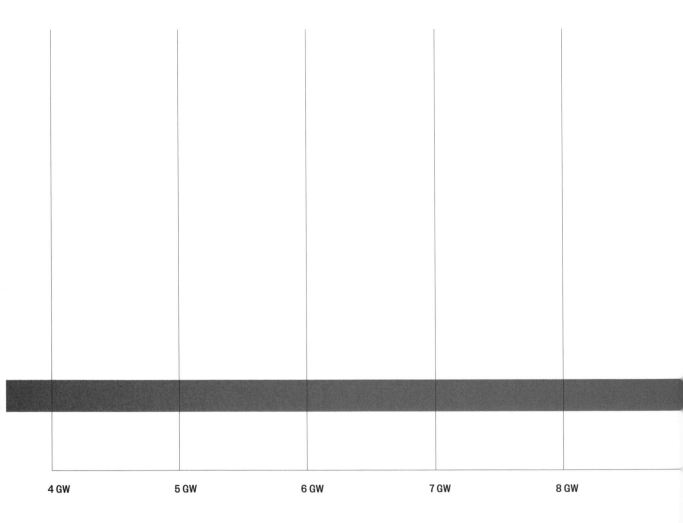

4 GW   5 GW   6 GW   7 GW   8 GW

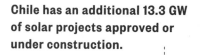

Chile has an additional 13.3 GW of solar projects approved or under construction.

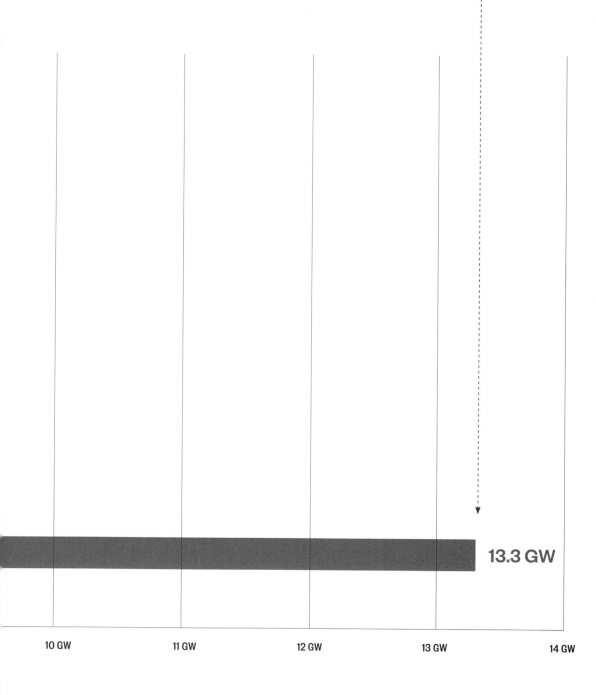

13.3 GW

10 GW   11 GW   12 GW   13 GW   14 GW

Batteries: The Key to a Renewable Future

MODERN CIVILIZATION DEPENDS upon a constant, reliable stream of energy. However, renewables such as wind and solar are notoriously intermittent; wind depends on the whim of nature, and solar power dries up as the sun goes down. Batteries solve this problem by storing excess power generated throughout the day and supplying it in the absence of sunlight or wind. In addition, batteries respond well to high electricity demands, help lower energy costs, and ensure reliability. They are the most crucial components in any clean power future.

Power storage is a much more difficult technological problem than power generation. From lithium ion to rechargeable flow, inventors and developers have experimented with many new ideas. There is not yet a magic bullet to solve our power storing needs. The good news, however, is that in the past decade, batteries have made great strides in capacity and lower prices. This is due in part to the electric vehicle industry, which relies heavily on efficient lithium ion batteries.

In 2016, Tesla Inc. began manufacturing its Powerwall and Powerpack energy products at its Gigafactory, currently the world's largest lithium ion battery factory. The goal of the plant is to drive down the cost of the company's electric vehicle and energy storage batteries while also spurring innovation. Doing so, according to the company, will make renewable energy storage a more accessible and viable option.

Others have also begun to find ways to make batteries cheaper and more efficient. In February 2017, researchers at Stanford University made batteries from aluminum, graphite, and a substance found in mammal urine and fertilizer, urea. The urea provides an inexpensive, safe electrolyte, or ion source, for the battery. Researchers have said the rechargeable battery is ideal for renewable energy use because it is cheap, long-lasting, safe, nonflammable, and efficient.

As we attempt to stabilize humanity's troubled relationship with Earth, batteries have the potential and the strong likelihood of being the key to this sustainable future. ◉

The Tesla Powerwall, a residential, rechargeable battery designed to store renewable energy.

Lyndon Rive

Cofounder and CEO of SolarCity

SAN MATEO, CALIFORNIA

IN 2004, WHEN LYNDON RIVE was driving with his cousin Elon Musk to the Burning Man festival in Nevada's Black Rock Desert, Musk made an offhand suggestion for a national photovoltaic brand that would target the full spectrum of potential solar users, from homeowners to commercial properties.

Rive was immediately taken with the concept. "Humanity cannot survive anymore by burning fossil fuels," he told me. "And solar panels were coming down in price. It was this inflection point where a solar business was finally feasible."

The idea was also appealing because Rive had been conscious of the climate crisis for years. He's an avid ocean diver, so he'd often seen the environmental fallout of carbon emissions firsthand.

"I would go to an amazing reef, come back a few

SolarCity employees install solar panels on the roof of a home.

years later, and notice that it's far more dead," said Rive. In addition to the ocean becoming warmer, Rive pointed out, "The water becomes acidic. It's, like, burning and dying."

As he explored Musk's idea, Rive came to believe the photovoltaic industry was ripe for disruption. The market was highly fragmented, and existing ventures had struggled to establish a recognizable brand. Worse yet, customers reported uneven experiences with solar installations.

When Rive founded SolarCity, he encountered resistance from Silicon Valley elites who tended to shrink from high labor business models. But SolarCity, which was acquired by Musk's automotive venture, Tesla, in 2016, soon exploded in popularity. It has attracted hundreds of thousands of household customers—each of whom on average will offset an estimated 178 tons of carbon over 30 years.

"Over the next 10 years, there's going to be a massive transformation from fossil fuel–based energy to renewable power," Rive said.

That's the reason he remains bullish on the future of the solar industry: it's a solid investment.

"Given the choice to pay more for dirty energy or less for clean energy, customers choose to pay less for clean energy," Rive said. "It's about independence, it's about security, it's about owning your own stake in renewables." ◉

There is other good news, too. Looking at this chart of CO₂ emissions, we can see that global emissions have stayed flat for three years in a row.

GLOBAL CO₂ EMISSIONS

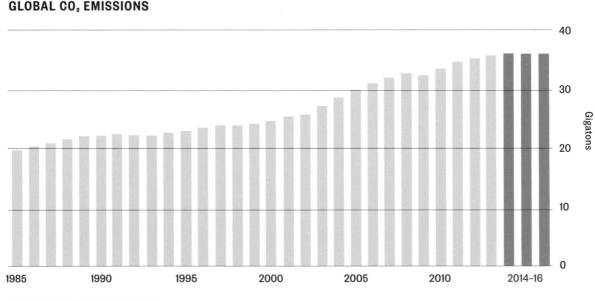

Gigatons

1985   1990   1995   2000   2005   2010   2014–16

SOURCE: GLOBAL CARBON PROJECT

Charging stations like this one power electric vehicles that do not emit CO_2.

Mountain View, California
August 24, 2016

We have an obligation to accelerate the move-ment toward meaning-ful changes in policies in every nation on the face of this Earth to stop the destruction of the global ecological system.

These are all of the coal plants that were proposed in the past 10 years and have been defeated, all the existing coal plants that were retired, and all the coal plants where retirement has been announced. All of these coal plants have been canceled.

We are getting off of coal.

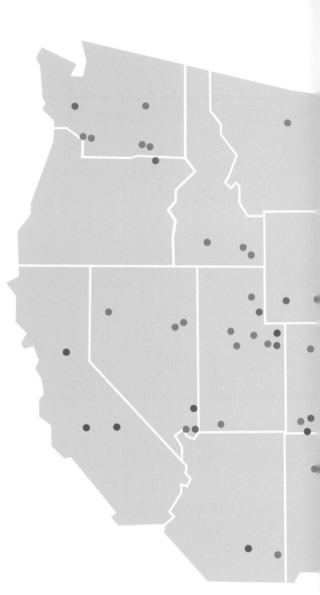

THE U.S. IS MOVING AWAY FROM COAL

- Retired
- Retirement proposed
- Proposed and canceled

SOURCE: THE SIERRA CLUB

  TRUTH TO POWER

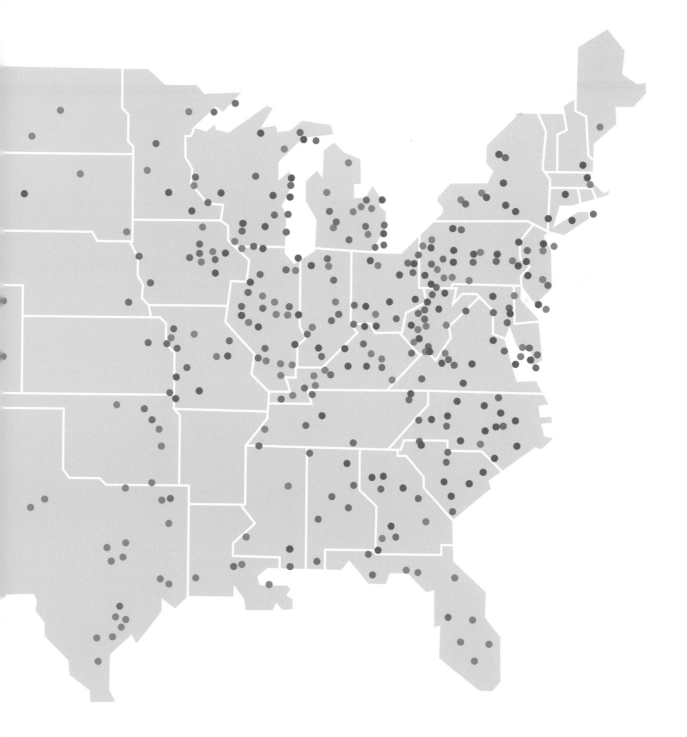

Bjarke Ingels

Architect and Founder of Bjarke Ingels Group (BIG)

COPENHAGEN, DENMARK

WHEN BIG SET OUT TO DESIGN a new waste-to-energy plant in Copenhagen, the company saw it as an opportunity to change public perception of what a utility building could be. Copenhagen is a flat city, so the group decided to embrace the structure's 85-meter height by giving it a sloped roof, complete with trees and trail markers. It would double as a ski slope during Denmark's snowy winters.

Bjarke Ingels, who heads the renowned Bjarke Ingels Group, also devised a "chimney" for the low-emission plant in collaboration with German architecture firm Realities:United that, instead of producing smoke, will puff out gigantic rings of steam each time the plant emits a metric ton of CO_2.

"There are people who imagine that the world is going down the drain," Ingels said, "and there are people who are optimistic about growth, that technology will provide new tools that will save us from problems."

Ingels belongs to the latter category. It's a design philosophy he calls " 'hedonistic sustainability,' which is the idea that sustainability is not a moral burden or sacrifice but a way to improve the quality of life and human enjoyment."

Ingels has made his career transforming cityscapes, but he was born and raised on a lakefront north of Copenhagen. That natural setting made a permanent impression on him, and since he was a young student, he has been drawn to questions of environmental policy and sustainability.

In what might be one of his most innovative projects to date, Ingels, together with the London-based Heatherwick Studio, designed a new headquarters for tech giant Google that features a canopy of photovoltaic panels that can be adjusted to let the perfect amount of sunlight shine down into the offices below while using the rest to generate electricity.

"We harvest every single photon," said Ingels, who often sports messy hair and a boyish grin. The idea behind the Google project, he said, was to use "existing technologies in ways that take advantage of all available resources."

Ingels believes that innovation is an inexhaustible resource—and that's what gives him hope for the future of the planet.

"There have been a handful of revolutions where, when humanity has a vision and a goal and a consensus about it, we can deliver striking results incredibly fast," he said. "All it takes is the resolve and the sense of urgency." ◉

An artist's rendering of the Amager Bakke waste-to-energy plant.

Copenhagen, Denmark

Ontario, Canada got off of coal as a matter of public policy.

The Nanticoke Generating Station was the largest coal-fired power plant in North America.

Nanticoke, Canada
July 26, 2009

The province was the first in North America to shut down its entire coal operation. Its last coal power plant closed down in 2014; a decade earlier, 25 percent of Ontario's power came from coal. One of the worst polluters in Canada, the Nanticoke plant site, is being repurposed as a solar farm.

Why Are Satellites Vital to Fight Climate Change?

WHEN THE CREW OF APOLLO 17 sent back a photo of the entire Earth in one shot in 1972, it was the first time we had seen the *Blue Marble* of our planet in all its majesty.

Today, satellites are one of the only ways to see Earth from afar, where we can observe climatic, weather, and energy patterns. However, even satellites can be limited in their scope.

When I saw that photograph of our planet, I was in awe. I thought, what if we could have these images on a daily basis? What if these images could be more comprehensive and detailed? Could they help build commitment to solving the climate crisis?

As I soon learned, there was nothing at the time capable of delivering what I had hoped for.

During my time as vice president, and inspired by the *Blue Marble* photograph, I called up NASA and proposed we send out a satellite that could get a broader view of Earth, and could stream the view in real time. I hoped that a fresh look at the entirety of our planet could inspire a new generation of environmental activists.

The result of the subsequent work was the Deep Space Climate Observatory satellite (DSCOVR), which to this day provides a full-sphere view of the Earth—and also constantly monitors the sun for potentially dangerous solar activity and solar storms that could threaten electric utility grids and pipelines.

In 2000, DSCOVR, which at that time was named Triana, after the first crew member on Columbus' ships to sight the New World, was approved by Congress despite some opposition. I was about to run for president, and that may have had something to do with the pushback.

In 2001, the satellite was built and ready to go when the new Bush-Cheney administration canceled its launch. At that time, DSCOVR's mission had become even more important as the older satellite that monitored solar activity was aging and carried dead instruments.

When the new administration canceled DSCOVR's launch, it didn't realize it was also cancelling the important solar-storm-monitoring warning system. Excessive energy released during solar storms on the sun can disrupt power grids, telecommunications, and GPS here on Earth. The warning system gives a 15- to 60-minute alert, allowing industries that depend on these technologies to prepare themselves for any disruptions.

Industries that could be negatively affected without the solar-storm warning system began to speak out against the decision to cancel DSCOVR, prompting the Bush-Cheney administration to agree to launch DSCOVR. But the administration proposed removing all of the climate-monitoring instruments, as well as the camera, and replacing them with the equivalent of sandbags. *Wow*, I thought, *that is extremism*.

DSCOVR gathered dust in storage for the remainder of Bush and Cheney's time in office.

Under the Obama administration, the satellite was retrieved from storage, tested, and found to be ready. DSCOVR was launched in February 2015. It now sits a million miles away from Earth at the L1 Lagrange Point, a spot between the Earth and the sun where gravitational and centrifugal forces are in balance and a satellite can remain in equilibrium,

The DSCOVR satellite sometimes captures an image of the Moon passing between itself and the Earth.

orbiting the sun in tandem with the Earth.

Onboard the satellite are three instruments, that work with information from other satellites to provide the full-sphere view of Earth as well as accurate information about our ozone, dust, cloud cover, vegetation, and volcanic ash—all important aspects in our study of climate change. Most significantly, DSCOVR provides for the first time a complete energy budget for the Earth, and monitors how much of the planet's energy from the sun is sent back out into space.

All these elements are what led to the idea of the DSCOVR satellite. Not just the desire for full pictures of our planet and our home, but also the amazing scientific data gathering that can be achieved from that special point in space. We had a real opportunity in the 1990s to start building enough public support to really get on track to solving the climate crisis. We lost that opportunity and we cannot afford to lose it again.

Just like Apollo 17 did several decades ago with the *Blue Marble* photograph, DSCOVR and satellites like it give us the means of getting an accurate view of our planet, and the data and inspiration we need to protect it. Let's use it. ◉

DSCOVR'S POSITION RELATIVE TO THE EARTH

Engineers prepare to launch the DSCOVR satellite.

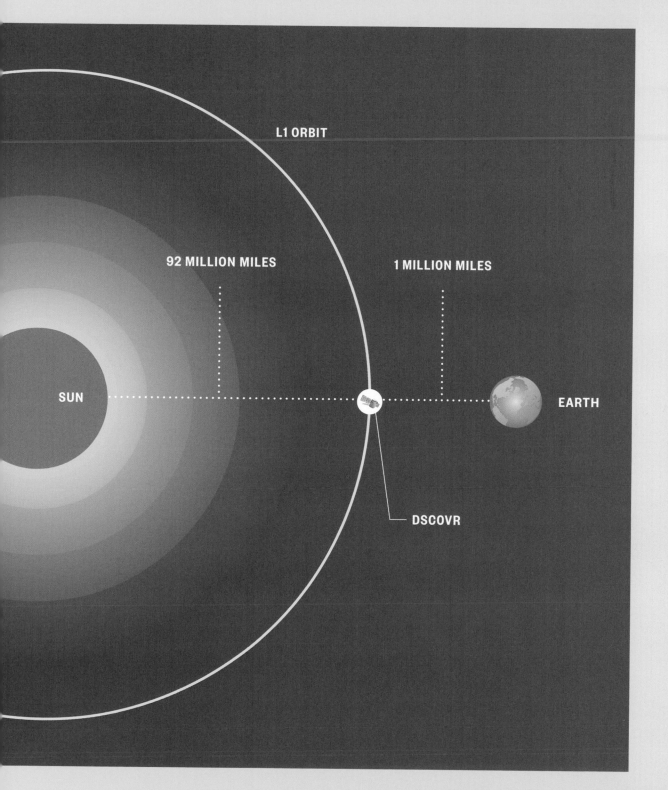

L1 ORBIT

92 MILLION MILES

1 MILLION MILES

SUN

EARTH

DSCOVR

In December 2015, representatives of virtually every nation in the world gathered in Paris to address the climate crisis. They all pledged to work together to achieve net zero greenhouse gas emissions as early in the second half of this century as possible. This was a historic agreement.

I was relieved when I got to Paris because there were influential men and women from countries around the world with whom I'd had the privilege of working, some of whom had been through my training program. They asked me to help make the conference a success.

To celebrate the Paris Agreement's entering into force, the Eiffel Tower is lit up with the words "The Paris Accord is done."

Paris, France
November 4, 2016

What is the Paris Climate Agreement?

195 countries signed a pledge to keep global temperature rise below 2°C (3.6°F), and, if possible, below 1.5°C (2.7°F).

▶ All countries agree to reduce global greenhouse gas emissions to net zero as soon as possible in the second half of the century.

▶ The U.S. pledged to reduce greenhouse gas emissions by 26 to 28 percent below 2005 levels by 2025.

▶ India aims to install 175 gigawatts of renewable energy capacity by 2022.

▶ China will peak its CO_2 emissions by 2030.

▶ Developed countries will provide $100 billion in climate finance by 2020.

▶ Countries should raise the ambition of their initial commitments over time to make sure we meet the goals of the Paris Agreement.

▶ The Paris Agreement entered into force on November 4, 2016.

Celebrating the successful
conclusion of negotiations.

Le Bourget, Paris, France
December 12, 2015

Christiana Figueres

Former Executive Secretary of the United Nations
Framework Convention on Climate Change (UNFCCC)
and a Climate Reality Leader

SAN JOSÉ, COSTA RICA

IN THE EARLY 1990S, when Christiana Figueres's daughters were young, she became preoccupied by the extinction of *Incilius periglenes*, better known as the golden toad. The golden toad had been a beautiful, bright-orange species native to the forests of Figueres's home country, Costa Rica, and its loss seemed to her emblematic of the deteriorating Earth she would someday pass on to her children.

"I was profoundly impacted by having witnessed the disappearance of a species," she said. "It was then that I started learning about climate change, and I have been passionate about it ever since."

Figueres had been thinking about sustainable lifestyles for a long time. Years earlier, before she studied social anthropology at the London School of Economics, her undergraduate work at Swarthmore College in Pennsylvania brought her to a remote Costa Rican village in the Talamanca Mountains, where she spent time with the indigenous residents. The experience, working with people who had no electricity or running water, permanently informed her outlook on the environment.

"Indigenous populations are some of the most vulnerable populations to climate change, but they are also treasure troves of responsible land use and water management practices," she said. "We can all learn from their wisdom and long-standing experience. We can all gather insights from them on the natural resource balance which we have destroyed and must reestablish."

But it was the episode with the golden toad that galvanized Figueres, whose mother had been an ambassador and legislator, and whose father served as the president of Costa Rica for three periods between the 1940s and the 1970s. In her parents' footsteps, she set out to become a leader for environmental progress.

To do so, Figueres chose a life of diplomacy. In 1995, she joined Costa Rica's climate change negotiating team at the United Nations Framework Convention on Climate Change (UNFCCC)—the group tasked, essentially, with negotiating a way for the world to avoid environmental collapse. Simultaneously, she founded the Center for Sustainable Development of the Americas, a nonprofit that advocated for financial systems that would promote sustainable development in the Caribbean and Latin America.

In 2007, Figueres attended the second-ever Climate Reality Training in Nashville, Tennessee. In 2010, she was appointed to run the Secretariat, the group that runs the UNFCCC. In that position, Figueres has acquired a reputation as a dynamo—skilled in the art of diplomacy but willing to push hard to convince the group's 195 member nations that the climate crisis is genuine and immediate.

The now-extinct golden toad.

"I'm very comfortable with the word 'revolution,'" Figueres told *The New Yorker* in 2015. "In my experience, revolutions have been very positive."

A key component of Figueres's strategy in that position has been empathy. She understands that Costa Rica's hydropower and wind resources give it an incentive to promote climate progress that not every country enjoys—especially not those with significant fossil fuel resources.

Figueres is known as an avid distance runner and dancer who tends to take public transit to work. In the wake of the Paris Agreement, she has set her sights on convincing world leaders to seize control of climate emissions by 2020, a year that scientists see as the deadline to accelerate CO_2 reductions in order to protect the world's most vulnerable populations from the worst impacts of climate change.

"We have not reached the turning point of greenhouse gas emissions," she said. "Industrialized countries must continue to decrease their emissions."

In sum, she sees the global community as having earned a mixed report card over the past 25 years. On one hand, she sees the implementation of the Kyoto Protocol and the Paris Agreement and the substantial investments and price reductions in renewable technologies as triumphs. On the other, she worries that the slowing of the temperature increase hasn't been enough to avoid the worst of the global climate change fallout.

Remembering the golden toad, she says, she draws hope from the legacy of the Earth that she'll leave behind for her children.

"We cannot afford the luxury of not being hopeful," she said. "This is a challenge we must face and conquer. Succeeding in arresting climate change is the only option we have. And success always starts with optimism." ◉

Even though President Trump has slashed federal programs to reduce emissions, U.S. businesses are moving ahead on their own. Many cities—and 36 states—have set renewable energy portfolio standards or goals.

CHANGE IN CALIFORNIA GDP AND GHG EMISSIONS SINCE 2000

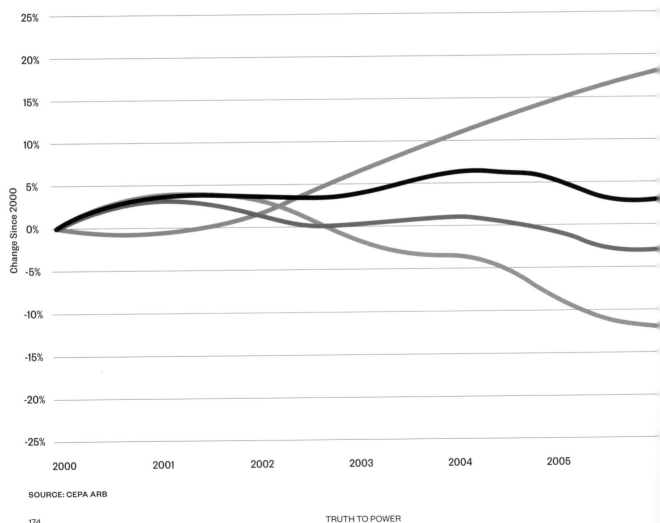

SOURCE: CEPA ARB

Their actions show that economies can reduce emissions and still grow. California leads the way— as its economy and population have grown, its greenhouse gas emissions have fallen. The state has pledged to reduce emissions by 40 percent by 2030, compared to its 1990 levels.

■ GDP
■ GHG EMISSIONS
■ GHG EMISSIONS PER CAPITA
■ GHG EMISSIONS PER GDP

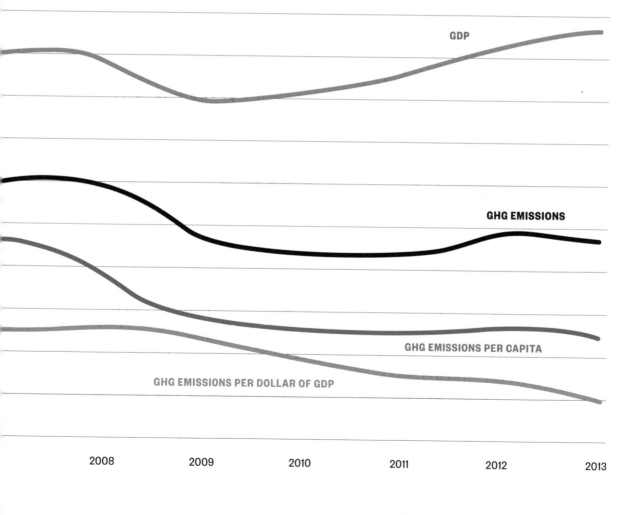

GDP

GHG EMISSIONS

GHG EMISSIONS PER CAPITA

GHG EMISSIONS PER DOLLAR OF GDP

2008   2009   2010   2011   2012   2013

POWER

This section presents a blueprint for what you can do personally to hasten the solution to the climate crisis. **As you may have noticed, the year 2017 has already been marked by an enormous upsurge in political activism in the United States, especially on the part of the many millions of Americans who strongly oppose the policies and proposals of the Trump administration.**

SPEAKING TRUTH TO POWER

The first part of this book presented the evidence for why it is so important that we quickly shift from dirty, highly polluting, fossil fuel–powered electricity generation to renewable sources of energy. The existential threat we are facing also makes it important that we shift to sustainable agriculture, sustainable forestry, electric vehicles, LEDs, efficient and affordable energy storage, and the introduction of hyper-efficiency in the way we use energy and materials. But in order to accomplish this transformation of human civilization in time to avoid the catastrophic consequences of the climate crisis, it is imperative that individual citizens become actively engaged in the struggle for the future of humanity.

Political will is a renewable resource, but it can only be renewed with the passionate involvement of individuals around the world who are willing to put time and energy toward learning the best ways to encourage political and business leaders at every level of society

It is imperative that individual citizens become actively engaged in the struggle for the future of humanity.

to make these changes a priority—and then take action.

This section of the book is designed to be a "how to" guide for those who want to become effective advocates for solving the climate crisis. It includes sections on how to encourage others to vote, how to most effectively influence the elected officials that represent you at the local, state, and national level, how to translate your passion into a successful strategy for changing minds and hearts—and how to persuasively press for the changing of laws and policies.

There continues to be a lot of resistance to doing the right thing, and it is important to understand the reasons so many are still reluctant to accept the urgency of the changes we must make. For starters, many people choose inaction because it is simply easier to embrace any available doubt that

Demonstrators from around the globe demand world leaders take climate change seriously.

New York, New York
September 21, 2014

serious change is truly necessary.

Unfortunately, powerful and wealthy special interests whose business models would be severely harmed by a speedy reduction in our dependence on fossil fuels have a sophisticated understanding of our vulnerability to this doubt. And in order to exploit that vulnerability, they have spent enormous sums of money to sow confusion by distributing an endless stream of falsehoods about the crisis and about the availability of solutions.

Scholars have documented the cynical process by which the carbon polluters have pursued this massive disinformation campaign. We can see with great granularity how the deniers have adapted the blueprint from a similar effort decades earlier by the cigarette manufacturers who, for 40 years, successfully delayed recognition of the scientific and medical consensus linking smoking to lung cancer and other diseases of the lungs and heart.

In their ongoing effort to cultivate paralyzing doubt about the reality of the growing disruption of the Earth's climate pattern, carbon polluters have also developed a robust alliance with ideologically motivated advocates for radical reductions in the role of government at all levels—or, as President Donald Trump's controversial associate Stephen Bannon said in early 2017, "the deconstruction of the administrative state."

Ironically, special interests have such an unhealthy degree of influence over governments at every level that the "administrative state" now subsidizes the burning of fossil fuels in the U.S. at a rate 25 times larger than the meager subsidies for renewables, and often impedes the transition to a sustainable future. As a result, the changes we need will not necessarily result in an increase in the role of government at all, but require instead a redefinition and redirection of the role government plays.

In order to bring about the changes we need, activists need to focus not only on communicating the truth about the climate crisis and the readily available solutions, but they must also focus on learning how to wield political power— the healthy and liberating form of power that democracy puts in the hands and hearts of every citizen who wants to exercise it.

Take, for example, the laws now in place in the state of Florida that were written by lobbyists for the powerful

Arizona Senator Jeff Flake answers questions at a town hall while the audience holds up red cards in disagreement.

Mesa, Arizona
April 13, 2017

fossil fuel–burning utilities there. At the present time, if an individual homeowner wishes to lease solar panels for his or her rooftop, Florida's state government forces them under power of law to abandon any such dream—unless they contact the fossil fuel–burning utility that serves their neighborhood, asks them for permission to lease a solar panel, and then pays money to the utility. If citizens of Florida can find a way to eliminate that onerous and ridiculous law, they will diminish the role of government, rather than increasing it. It is, again, a question of power. Who has the power? Voters or special interests?

Recently, when Floridians eager to generate their own electricity from their own rooftops—and exert their independence from the monopoly power of the utilities—complained that Florida is, after all, the "Sunshine State," the head of one of the two largest utilities replied, "We are the Sunshine State, but we're also the partly cloudy state."

American democracy has been hacked. Elected representatives of the people have too often become obedient, obsequious servants of the utility companies. At the same time, the utilities also spend large sums of money to flood the airwaves with deceptive television and radio advertisements intended to brainwash viewers into passivity and fool them into thinking that the utilities have their best interests at heart—when actually, the utilities are simply using falsehoods to increase their profits and executive bonuses.

There are many other examples of what is happening to Florida throughout the United States and in many nations around the world. And tragically, many legislators have learned that if they stay on the "gravy train" provided by special interests, they can rely on those interests to fool voters into thinking that they are being well represented, and persuade them to reelect their dishonest representatives over and over again—even if those representatives never lift a finger to serve the public interest.

However, in the dawning age of the Internet and social media, individual voters are beginning to awaken to the way special interests have taken advantage of the general public. As a result, we are now witnessing the beginnings of a nonviolent political revolution in which

Rex Tillerson, former CEO of ExxonMobil, after being sworn in as Secretary of State.

Washington, D.C.
February 1, 2017

American democracy has been hacked.

citizens are starting to take back control of their own destiny and are demanding that those they elect to serve the public interest actually do so.

If you want to join this revolution—and I hope you do—what follows is a practical guide to how you can save the future. As you will see, different people find different paths. Each of us has our own areas of interest and we all naturally choose different ways to help win the battle for our future.

Once you make the basic decision to get involved, you will discover the best ways that you can be effective. But remember, as Goethe once wrote, "It is not enough to know; we must also apply; it is not enough to will, we must also do." The truth can set us free, and once free, we must act. ◉

Be an Involved Citizen

There are a number of ways individuals can have a surprisingly big impact on elected officials at every level of government. I used to be an elected official, and I know a little bit about what works. The secret is to make it unmistakably clear that their positions and actions related to solving the climate crisis will absolutely determine whether you (and everyone you can influence) will either:

A strongly support them for election or reelection

B do everything in your power to make certain they are defeated in the next election.

Delegates voting to nominate
candidates for the 2016 election.

—

Las Vegas, Nevada
June 28, 2016

1. Register to vote and help get others to vote too.

Over 40 percent of the eligible voting population did not vote in the 2016 presidential election. Stay registered no matter what—every election counts.

▶ Find out how you can register at usa.gov/register-to-vote.

Encourage all of your friends and other people in your social network to register, and then make sure they vote on election day. You can consult resources like Rock the Vote, Project Vote, NextGen Climate, and League of Conservation Voters to learn more about getting out the vote.

2. Find your elected officials.

In addition to the president, you are represented by members of the U.S. House and Senate at the federal level, your governor and state legislators at the state level, and your mayor and local officials in your city and county.

▶ To learn who represents you at each level, visit usa.gov/elected-officials.

3. Learn how they vote.

It's important to understand how your representatives vote on climate issues. This will help you better understand their commitments and dictate how you engage them.

▶ Visit online voting scorecards such as Scorecard.LCV.org or GovTrack.us.

4. Call, tweet, and write to your elected officials.

Once you've learned the names of your lawmakers, you can reach them through the congressional switchboard by calling (202) 225-3121 and telling the operator to whom you wish to be connected. Save this number in your phone so that you can call them regularly.

Calling members of Congress is one of the most effective things you can do as an engaged citizen. Congressional staffers will tell you how important these calls are in helping lawmakers understand where their constituents stand. The number of calls an office receives has a significant impact on how a representative votes on a piece of legislation. With that said, any sort of contact with your officials strengthens our democracy.

Republican Congressman Rodney Davis now supports finding solutions for the climate crisis.

Springfield, Illinois
August 14, 2014

Call your elected officials.

Very rarely, you will be able to talk directly to an elected official, but it's most likely that you will speak to a staff person. When calling, it's important to let them know exactly what issue you're calling about and concisely tell them your position. Most of the time the staffer will simply be recording the number of calls they receive on a given issue, so there is no need to speak at length and you shouldn't feel pressured to carry on a conversation. Be polite, but by all means be forceful and make it unmistakably clear that you are passionate about this issue and that your network of friends will be tuned in to exactly what the elected official does or does not do relevant to climate. And get others to call too. Numbers matter.

SAMPLE CALL SCRIPT

TO CALL REGARDING A GENERAL ISSUE

"Hello, my name is [YOUR NAME], from [YOUR CITY], and I'm calling to urge Representative/Senator/Governor/Mayor etc. [   ] to advocate for genuine solutions to climate change. This is important to me because [VERY SHORT REASON]."

TO CALL REGARDING A VOTE ON A BILL

"Hello, my name is [YOUR NAME], from [YOUR CITY]. I'm a constituent and I'm calling to urge Representative/Senator/Governor/Mayor etc. [   ] to vote yes/no on bill [####]. This is important to me because [VERY SHORT REASON]."

A man holding a ballot gets ready to vote.

—

Cottonwood Falls, Kansas
November 8, 2016

Tweet your elected officials.
Members of Congress and other elected officials also pay careful attention to replies and mentions on their Twitter accounts, especially when they come from people who clearly identify themselves as constituents. One study showed that even a few dozen tweets from constituents on a given issue was enough to get the office's attention. ◉

SAMPLE TWEET

Dear @[YOUR OFFICIAL'S TWITTER HANDLE], I'm a constituent. Please vote NO on any bill that would lead to more burning of dirty coal.

See page 204 for more about using social media as a climate activist.

Senator Lamar Alexander
455 Dirksen Senate Office Building
Washington, DC 20510

July 28, 2017

Clearly state the bill name or number you are writing about in the first sentence

Dear Senator Alexander,

Identify yourself as a constituent and include relevant information about yourself, like profession, if you're a parent, veteran, etc.

I'm writing to urge you to support the Global Warming Solutions Act of 2017, which takes an important step toward reducing carbon emissions that cause climate change.

As a constituent and parent of young children, I am very concerned about the future of our planet.

NASA recently confirmed that the Earth has set record high temperatures for three years in a row. There is no time to waste. Congress must take bold action to hold corporations accountable and move toward renewable energy sources, which will be a boon to our economy as well as our environment.

Tie it to current news

We are making progress to solve the climate crisis, but our political leaders must act. Please take the next step by voting yes on the Global Warming Solutions Act.

Sincerely,
Jefferson Smith
350 Green Terrace
Nashville, TN 37221

Make sure your voice is heard.

Win the conversation.

A woman asks a question of Senator Lindsey Graham at a town hall meeting.

———

Columbia, South Carolina
March 25, 2017

Speak at a Town Hall Meeting or Forum

One of the most effective ways to influence your elected officials is to speak to them face-to-face in town hall meetings or at other open forums. Members of Congress and local elected leaders want to know which issues are most important to their constituents, and taking the time to show up in person communicates to them that you are serious about addressing climate change at the municipal, state, and federal level. Even if you're a little nervous about public speaking, with a little preparation you can make a big impact. This is what speaking truth to power is all about.

1. Make your concern about the climate personal.

Successful climate activists are experts at connecting their personal stories and concerns to the greater issues of climate change. It's important you pick an issue that affects you and your community personally.

Here are some ideas:
▶ Promote clean air by asking that municipal vehicles be transitioned to electric vehicles or hybrids.
▶ Make sure your tax money isn't contributing to climate change by asking that municipal buildings transition to solar or wind power.

Paying attention to the climate and environmental issues being discussed in your community and at the state and national levels may provide some good ideas for how you can get involved.

One example is The Climate Reality Project's 100% Committed campaign: encourage your city, business, or university to commit to using 100 percent renewable electricity.

▶ Learn more at climaterealityproject .org/content/roadmap-100.

Governments everywhere should put a price on carbon that reflects the true costs we all pay for global warming pollution and incentivizes the transition to a clean energy economy.

▶ To learn more about this, visit the Climate Reality website at climaterealityproject.org for best practices on designing such a policy.

You can also focus on related issues, but remember to mention climate early in your remarks.

QUICK TIP

If your city doesn't have a climate task force, maybe it's time to form one! *See page 252 for information on running for office.*

2. Find the right opportunities to speak and engage with elected officials.

Organize or join local groups, like the Indivisible chapter in your community (see page 284), to engage the political system at town halls, public events, district offices, rallies, marches, and relevant city council or committee meetings. See if there is a specific municipal council or body focused on issues related to energy and the climate. Find out if you need to do anything in advance to speak at these events or just show up ready to contribute.

3. Prepare to speak.

Remember that in most town hall meetings and local forums like city council meetings, speakers are typically expected to speak for only a few minutes, so make your point quickly and effectively. In order to have the most impact, you may want to consider doing the following to prepare:

▶ Plan out your remarks ahead of time, and write them down to help commit them to memory. Make your most important point early in your remarks.
▶ Practice in front of friends or family to get feedback, and time yourself to make sure you're within the time constraints.
▶ It's important the committee knows you're an active member of the community. If you're a business owner, parent, veteran, faith leader, medical professional, or educator, make sure to include this information in your presentation.
▶ Bring a visual aid or two, whether it's a prop or a poster board with some key facts or graphs, to help you make your point.

▶ You want to feel confident and make a good impression. You don't need to wear a suit, but dress professionally and avoid wearing political T-shirts or buttons unrelated to the cause you are speaking about.

4. Bring backup. There's strength in numbers.

In order to win over your officials, you've got to persuade them that this issue is not only important to you, but to many others in your community. If possible, it's a good idea to bring friends and family to show support for your presentation. Don't be shy about asking them to clap or vocally agree during and immediately following your presentation.

5. Then, when the time is right, speak up!

There's nothing to worry about. As a climate activist, you know the science is on your side. Arrive early and take a seat somewhere with easy access to the aisle. When your name is called or when your raised hand is called upon, take a deep breath, and give your presentation. The more you speak up, the more comfortable you will get, and the more effective you will be. ◉

We ought to feel a sense of joy that we are alive at a moment when we can join with one another in a great cause, the stakes of which have never been higher.

THERE'S NO PLANET B

Demonstrators throw a giant
Earth balloon during the COP21
climate change meetings.

Rome, Italy
November 29, 2015

The Greenville News

Paris Agreement Can Grow U.S. Jobs

By Ryan Popple
CEO, Proterra – Guest

Two actions inside of the Trump White House are converging: First, President Trump has met with leading CEOs several times since his term began. Why? Because he wants to restore U.S. manufacturing jobs, a message he forcefully reiterated during his address to Congress in February. Second, Trump's recent Energy Independence executive order, while dialing back some of former President Obama's climate policies, did not specifically address the US's participation in the Paris Agreement. What do these two seemingly unconnected events have in common? Everything.

I'm the CEO of the leading manufacturer of electric buses, Proterra. We employ 150 talented South Carolinians to manufacture American-made, all-electric city transit buses that out-perform our Chinese competitors. Buses manufactured in Greenville are now serving mass transit routes throughout the country – riders in Nashville, Seattle, Louisville, Stockton, Los Angeles County, Park City, Tallahassee and Seneca, SC are all safely transported on clean, quiet Proterra buses. In 2016 alone, our six buses in South Carolina drove 175,000 miles while saving nearly 44,000 gallons of diesel. And with a backlog that totals nearly 300 buses, we've developed the future of the U.S. transit market. Within 10 years, I doubt any city in the U.S. will buy a fossil-fuel transit bus. EV is just too efficient, too good, too affordable. South Carolina should be very proud that the EV transit trend started right here, with great engineers and manufacturing.

In addition to the 150 people employed at our Greenville manufacturing facility, Proterra has helped to create hundreds of other manufacturing jobs indirectly in South Carolina. We are American-made, through and through. Proterra buys approximately 22% of all components from within South Carolina. If we expand to the entire South East, Proterra buys approximately 36% of all components locally. Nearly 80% of all of our components come from other American companies. Buy America is in our DNA.

We're not alone in this jobs effort. Newer and cleaner technologies are producing thousands of more U.S. jobs, including in manufacturing. Indeed, clean energy jobs are booming—employing now over 3 million Americans. According to one recent source, there are more jobs now in the renewable energy space in 41 states and Washington, DC than in the coal, oil and gas sectors. Over 100,000 U.S. workers now manufacture, construct and maintain the U.S. wind turbine fleet. The solar workforce increased by 25% in 2016, while wind employment increased by 32%. Manufacturing jobs within the solar industry rose 26% in 2016 and now total nearly 40,000 jobs. And projections are that the growing demand for electric vehicles will mean tens of thousands of more jobs for Americans.

A lot of our neighbor companies and institutions in South Carolina are already competing effectively in the global energy technology market that is expected to grow to $6 trillion. As an example, South Carolina is a leader in hybrid and electric vehicle production. The vehicles that BMW manufactures in Spartanburg, SC use advanced materials, state-of-the-art batteries and energy efficiency software to lower fuel consumption without reducing driving performance. And several firms manufacture components and equipment that are vital to the generation and distribution of clean energy.

More locally, the Upstate is a case study in why restoring manufacturing is the key to economic health. Thanks to forward-thinking leaders, the region decades ago invested in and attracted value-added manufacturing. The results have been phenomenal, and have led to a synergy among business, community and universities. We are proud to be part of this success story.

Unfortunately, outside of the clean energy sector, we have seen a tremendous loss of manufacturing jobs. As President Trump has noted, since 2000, the United States has lost more than 5 million manufacturing jobs and witnessed the closing of more than 60,000 factories. I agree with the President that we cannot afford to lose any more manufacturing jobs. But I disagree that more coal is the answer.

The Palmetto State agrees. In its September 2015 Final Report, the SC Clean Energy Industry Manufacturing Market Development Advisory Commission found that the State "is advantageously poised to capture a considerable portion of the growing clean energy manufacturing sector thanks to its skilled and growing manufacturing workforce, strategic location on the east coast and world renowned research infrastructure." Notably, the Report made clear that U.S. manufacturing and sustainability efforts were not mutually exclusive.

So what does any of this have to do with the Paris Agreement? It means that 190 countries have committed, in writing, to reducing their greenhouse gas emissions based on their own national plans. They will need to buy clean energy products (solar panels, wind turbines, electric vehicles and more) to meet their commitments under the Agreement—and those products should be made in the U.S. American companies like Proterra and thousands of others can employ millions of U.S. workers—including South Carolinians—to build advanced energy products and stamp them "Made in America" for export around the globe.

For all of these reasons, CEOs like me believe the U.S. should remain a part of the Paris Agreement, which involves action by all countries. I also am committed to increasing U.S. manufacturing jobs, in the Upstate and across our operations in the U.S. The two objectives are inextricably linked. Instead of allowing our global competitors to take advantage of the Paris Agreement, selling foreign products to countries that are trying to meet their emissions reductions goals, we should be out there competing for that work. Our foreign competitors are playing to win. We should be too.

(The piece originally appeared in The Greenville News, Sunday, April 2, 2017)

Write about Climate Effectively

By writing articles, opinion pieces, and letters to the editor, your reasonable, science-based arguments and personal stories can reach a massive audience and help counter misinformation about climate change—especially if your writing goes viral online. Influence and impact is within the reach of anyone who can write persuasively and passionately. Lawmakers, CEOs, and other leaders pay careful attention to the opinion pages of local and national papers, making them an important venue for persuading powerful people and affecting public discourse.

Even though special interests have more influence in our democracy than ever before, the opinions of voters can still carry the day—if enough people speak out passionately. You don't need any special qualifications or even a lot of experience to get published. Here are some best practices that will dramatically increase your chances.

1. Be timely.

The first question any opinion editor will ask when you submit a piece is, "Why is this relevant now?" With few exceptions, most pieces that are accepted are tied to a breaking news story or a hot-button issue that is generating discussion locally or globally. Look for opportunities to enter the conversation quickly, before the public focus moves on to the next topic. Pick a topic that you're passionate about and that you have something to add to. Here are some ideas:

► A local ballot measure, piece of legislation, or other policy issue.
► A personal story about what you are doing to reduce your own carbon footprint.
► An important racial, social, or economic justice issue that intersects with climate change and environmental protection.
► Criticism for elected officials who are failing to address climate issues or praise for those who are showing courage and leadership.

2. Be concise.

Research guidelines for letters to the editor and other pieces before submitting one. For an op-ed, 600 to 800 words

Even though special interests have more influence in our democracy than ever before, the opinions of voters can still carry the day— if enough people speak out passionately.

is standard, while letters to the editor are generally 100 to 300 words. Lengthy pieces may be dismissed outright, no matter how good your writing or argument. Some outlets accept longer-form pieces, but often not as unsolicited submissions. More important is your reader's attention span. You're more likely to reach a broad audience with a piece that is short, clear, and to the point.

QUICK TIP

Editors pay attention to how many readers a piece gets in the first couple of hours, so make sure to promote it extensively.

3. Support your argument.

Back up your points with concrete examples and evidence. Avoid hyperbole and exaggeration. If you state your points clearly and calmly while providing proof for your claims, your argument will be more convincing, especially for the most skeptical readers. It's also a good idea to refer to your personal experience and story.

4. Know your audience.

Put yourself in your audience's shoes and think about the arguments, sources, and tone that would be most convincing to them.

5. Pick the best outlet.

Choose your publication well. Which outlet will be most credible to your audience? Members of Congress, for example, are influenced by pieces appearing in local papers within their district as well as Beltway outlets like *The Hill* and *Politico*. *The Huffington Post* and *The Guardian* are well respected, but if you're reaching across the aisle, they may not be the best home for your op-ed. National papers like *The New York Times* and *The Washington Post* have a wide reach, but submitting to them is extremely competitive.

6. Get feedback.

Always try to get a second opinion before submitting your piece. If you're trying to reach a specific audience, get feedback from a member of that group. Even just having a friend or family member look over your piece can ensure an editor doesn't pass you over because of a simple spelling or grammatical error. If you don't have someone else who can review it, sleep on it and take a second look in the morning before you hit send. ◉

Start a Petition

The First Amendment says that no law shall prohibit the right of people "to petition the government for a redress of grievances." Nowadays, it has become common practice for people to unite in an expression of support or reproach for a government official or organization. Petitions can directly affect policies or signal to leaders how strongly constituents or customers feel about a particular issue. Every signature adds weight to the concern. If you're a budding climate activist, petitions are also a great way to build a community of like-minded people.

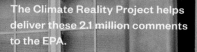

1. Identify your target.

Direct your petition to whoever has the most direct influence on the issue. For government concerns, this might be a representative voting on a bill. For corporate concerns, this might be a CEO, large shareholder, or the company's sustainability department.

2. Craft a compelling and concise message.

State the problem and desired outcome clearly, so everyone understands the "so what." Use short sentences and don't go over 200 words. Offer a deadline to create urgency and news clips to show the issue is a subject of public interest.

3. Circulate widely.

Your petition's success lies in getting it out there. Use social media networks to post and share the link. Share around your school or office as well. Bring paper petitions to meetings, or organize signature drives at local businesses, farmers' markets, and events.

4. Deliver.

If your petition is online, it is easy to bring your message to its recipient: hit send. Make sure your petitioners use their social media to follow up with the person or organization you are petitioning. If you have paper to deliver, mailing is an option. Or consider delivering it at an important event, like a shareholders meeting or a press conference.

Taken together, these steps can drive great momentum for your cause and your community. Keep your petitioners close, thank them for their participation, and constantly involve them in your future projects. ◉

Americans exercise their right to petition in opposition to a bank merger.

San Francisco, California
February 2, 2015

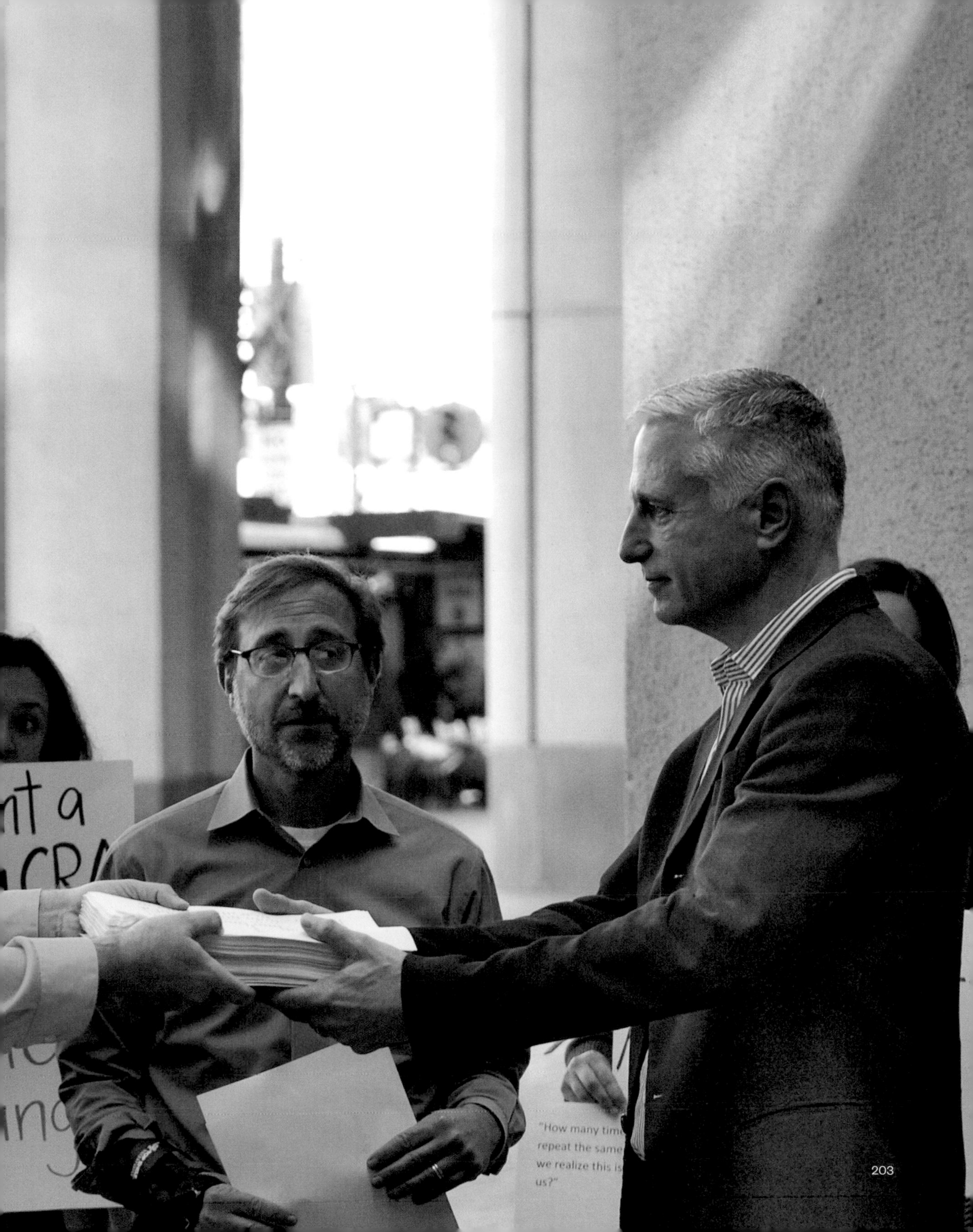
"How many time
repeat the same
we realize this is
us?"

Establish Yourself as a Climate Activist Online

The Internet has changed the face of activism. Some of the most popular modern political movements started online through social media, blog posts, and email campaigns. You don't need to rearrange your life to be a climate activist. You can start or join a community of climate activists around the globe from your sofa, and then work together to encourage good science and advocate for protecting the Earth.

Ed Stafford updates his blog during his record-breaking 859-day Amazon River walk.

Amazon Jungle, Peru

1. Create social media accounts.

Social media sites provide powerful platforms for change-makers to educate the general public and influence political leaders. Facebook, Twitter, Instagram, Snapchat, Pinterest, and Tumblr all have their value, but you may not have time to maintain a presence on all of them. Keep your finger on the pulse: different platforms rise and fall in popularity, and different demographic audiences tend to prefer certain platforms over others.

2. Choose your best outlets.

Online activism takes many forms in many venues these days. Social media is a necessity; beyond those platforms, you may also want to establish a blog, create an email newsletter or mailing list, publish on a writing platform such as Medium, or explore new outlets and platforms just launching.

3. Show that you're a climate activist.

As you set up your accounts, establish yourself as an outspoken climate activist by putting your climate advocacy front and center in your username, profile picture, and bio. You can identify yourself as part of the broader movement by using a profile image "filter" like the Climate Reality Project's green ring.

▶ Include a link in your bio to your blog, if you have one, or to an informative climate-related website like climaterealityproject.org.

4. Use strong visuals.

Especially when you're starting out, photographs, videos, infographics, and memes help grab people's attention and increase sharing.

5. Follow other influencers.

Pay attention to what prominent climate experts, political leaders, NGOs, and journalists are saying and amplify their messages or reply with a different opinion.

6. Be creative.

Humor, wit, and creativity can go a long way in helping you reach an audience beyond your immediate circles.

7. Share facts and breaking news.

Become a source where your followers can find important news and analysis on local and national climate issues.

8. Post frequently.

There are best practices for each platform as to how often you should post, but at least a few times a day is usually safe. During key moments like a State of the Union address or a vote in Congress, you should post several times using appropriate hashtags, as many more people will be paying attention. ◉

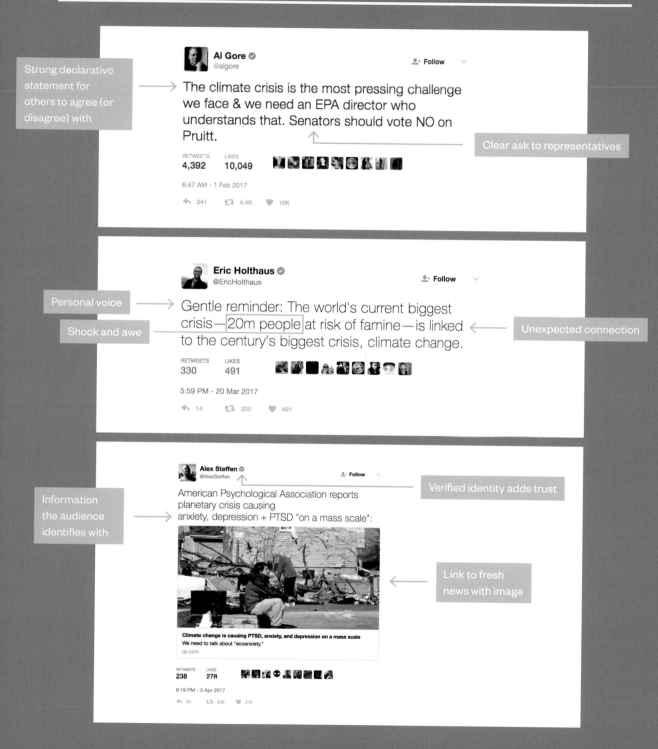

Strong declarative statement for others to agree (or disagree) with

Al Gore ✓
@algore

Follow

The climate crisis is the most pressing challenge we face & we need an EPA director who understands that. Senators should vote NO on Pruitt.

Clear ask to representatives

RETWEETS **4,392**   LIKES **10,049**

6:47 AM - 1 Feb 2017

241   4.4K   10K

Personal voice

Shock and awe

Eric Holthaus ✓
@EricHolthaus

Follow

Gentle reminder: The world's current biggest crisis—20m people at risk of famine—is linked to the century's biggest crisis, climate change.

Unexpected connection

RETWEETS **330**   LIKES **491**

5:59 PM - 20 Mar 2017

14   330   491

Information the audience identifies with

Alex Steffen ✓
@AlexSteffen

Follow

Verified identity adds trust

American Psychological Association reports planetary crisis causing anxiety, depression + PTSD "on a mass scale":

Link to fresh news with image

Climate change is causing PTSD, anxiety, and depression on a mass scale
We need to talk about "ecoanxiety."
qz.com

RETWEETS **238**   LIKES **278**

9:19 PM - 3 Apr 2017

20   238   278

John Cook

Founder of Skeptical Science

BRISBANE, AUSTRALIA

BY 2007, JOHN COOK had grown so frustrated by endless arguments with his family over the existence of climate change that he began to maintain a list of dubious talking points they used against him, complete with counterexamples he could use to cut the altercations short.

The technique was so successful that he started to wonder if his list might be useful to others. After some consideration, he turned it into a remarkable website called Skeptical Science, which assembles common myths about climate change. Think "Then why is it cold out?" and "There is no consensus"—alongside meticulously researched rebuttals.

The site touched a cultural nerve, and it was just beginning to pick up steam when Cook got an email that would permanently alter the course of his life. The message was from a cognitive scientist, who praised the site, but forwarded a selection of psychological research on how to construct an argument that's maximally effective at changing hearts and minds.

For Cook, it was an epiphany. He realized that there was an entire body of research that studied the question of how to communicate complex ideas most effectively. Embarrassingly, it soon became clear that he'd been making missteps that an expert in the field would consider rather basic.

"I was doing what almost everyone does when

they debunk myths," he said. "They emphasize the myth first. They usually use it as the headline, then talk about the facts later. What that does is make the myth more prominent, and it risks strengthening it."

Cook, who exudes a quiet intelligence, started to study with passion. At first he was just looking for tips to make entries on Skeptical Science more convincing, but he soon started to wonder if he might have found his calling. Eventually, he enrolled in a doctoral program at the University of Western Australia where he dove into intricate questions of communication theory—while continuing to maintain Skeptical Science on the side.

In the university system, Cook approached the challenge of communicating effectively with the same fervor he'd employed regarding climate change. He co-authored two books and penned a thesis about closing the "consensus gap" between the scientific case for climate change and public perception. He also served as the lead author on an influential paper, later tweeted by U.S. President Barack Obama, that quantified support for global warming among researchers.

Cook also developed a set of three guidelines that he believes are key to framing an argument as effectively as possible.

The first is to know your target readers. "The best audience is often not people who hold a different

Debunking a myth creates a gap

Fill the gap with a factual narrative

Cook's website features visual aids like this illustration to help people address climate denial.

view," he said. "It's the undecided majority. Trying to convince a hardcore disbeliever is unlikely to have much impact."

The second is to carefully consider the key facts you want to communicate and to boil them down to their absolute essence. Cook uses the phrase "simple and sticky" to describe facts that have been reduced to a form so pure that they cut through the spin and lodge in readers' memories.

To describe his third guideline, Cook uses an example from immunology. Cook has found that if you explain to someone why a faulty argument is weak, the effect can be akin to vaccination. When they eventually come across a stronger form of the obfuscation, they'll be better equipped to argue against it.

For instance, researchers in 2017 found that if they'd exposed subjects to the idea that politically motivated groups use misleading tactics to convince the public that there's wide disagreement among scientists about global warming, those subjects

became less likely to believe similar false claims in the future.

"You need to inoculate people against misinformation," he said. "It's about exposing them to a weak form of the misinformation, so that the strong form won't influence them."

If you want to promote climate science in the public sphere, Cook recommends that you follow his guidelines—and, hearkening back to the roots of Skeptical Science, suggests that you take the issue up with those closest to you.

"One of the most trusted sources of information about climate change is friends and family," he said. "I would suggest that people speak from the heart on why this issue is important to them." ◉

To learn more about Cook's work, visit SkepticalScience.com.

I believe we have the capacity, at moments of great challenge, to set aside the causes of distraction and rise to the challenge that history is presenting.

Magnify Your Impact with Press Coverage

Understanding the media landscape, including knowing the prominent climate publications and journalists, is vital for driving attention to your activism. Take note of how journalists cover various causes to get a feel for what piques their interest. Use this insight to organize creative, attention-grabbing campaigns that journalists will be unable to ignore.

1. Build relationships with journalists—it's easier than you think.

Send an email or letter to the reporters who cover the environment in your area, or even those at national outlets who cover the environment. Follow them on Twitter, comment on their articles, or send them a quick email to thank them for their reporting on an important issue. They'll be more likely to remember you later when you ask them to cover your event or petition.

2. Alert the press.

If you are planning some newsworthy statement or event, keep reporters informed and give them notice so they can plan ahead. Remember that journalists are inundated with story ideas, so sometimes a quick email with just the key facts is as effective as a formal press release. Don't pester, but if you don't hear back after sending an email or release, follow up with a phone call to make sure they saw it.

3. Prepare for the possibility that you might be interviewed.

What is the most important message that you want people to take away from a radio interview or article in which you are quoted? Write down three or four key talking points, rehearse them, and stick to them no matter what. If a reporter asks you a question to which you don't know the answer, or that seems off topic, find a way to transition back to your talking points. They should be no more than a sentence or two long. Avoid rambling.

Sometimes you will be dealing with a reporter who is hostile to your position. Maintain your composure and don't show anger or frustration. If you stay calm, you'll look like the reasonable one to anyone watching. ◉

A demonstrator is interviewed during a sit-in coordinated by the Campaign Against Climate Change.

London, England
March 7, 2015

Tais Gadea Lara

Journalist and
Climate Reality Leader

BUENOS AIRES, ARGENTINA

WHEN TAIS GADEA LARA was a young girl living near Buenos Aires, she told her mother that she intended to be a journalist when she grew up.

Gadea Lara, now 29, has emerged over the past decade as an award-winning environmental journalist in Argentina's independent media: she's the co-founder of media project Conexión Coral, a TV reporter for *Efecto Mariposa* and *Hoy Nos Toca*, and writes for newspapers and magazines including *La Nación* and *Sophia*.

Gadea Lara now lives in Buenos Aires proper, though when we spoke she had just returned from an expedition to report on conservationists in Patagonia who are working to protect the hooded grebe, a critically endangered migratory bird. It's the type of story that epitomizes her wide-ranging oeuvre, mixing a deep concern for sustainability and the environment with a hopeful view of the role the public can take in shaping the future of the planet.

"What I try to do in every article is to be optimistic," she said. "There are a lot of people working on climate action, and it's helpful to show readers that there are people in their communities who are working on that."

That outlook underlies Gadea Lara's ethos as a journalist. She reports on the grave realities of climate change, but tries to counterbalance each story with examples of governments and citizens that have taken action for the public good.

Last year, for instance, Gadea Lara visited vulnerable communities on the outskirts of Buenos Aires that lacked access to electricity or hot water. She saw that community groups had stepped in to provide photovoltaic panels and bathing facilities, and she ended up writing a newspaper feature about efforts in Argentina to fight poverty by harnessing renewable resources.

This positivity is key, she believes, to engaging citizens of a developing economy who might otherwise see environmentalism as a luxury for wealthier nations. It's the same viewpoint that appealed to Gadea Lara when she attended South America's first Climate Reality Leadership Corps training program in Rio de Janeiro in 2014—the idea that, if we work together, it's not too late to save the world.

"I think the main job we have as journalists," she said, "is to tell people what they can do." ◉

To learn more about Gadea Lara's work, visit her website at TaisGadeaLara.com.

Gadea Lara interviews a man about the impact of climate change on his life during the COP22 Climate Conference.

Marrakesh, Morocco
November 6, 2016

When children ask questions about what's being done to their planet, parents are often moved to make changes.

Children take part in a country-wide effort to break the world record for most number of people planting trees simultaneously.

Catequilla Hill, Ecuador
May 16, 2015

Talk to Children about Climate Change

When I was a young child, my father taught me about soil conservation on our family farm. For me, that was the beginning of my awareness of and concern about the impact people have on the environment. If we can help children make sense of the environmental problems they're inheriting, they'll be better equipped to solve them. Start early, with simple concepts, and make sure to nurture an ongoing conversation.

1. Foster a love of nature.

Children who care for nature are more likely to become adults who do too. Teach children to respect the Earth and take responsibility for keeping it healthy by getting them outside as much as possible.

WHAT HAPPENED TO PLAYING OUTDOORS?
Almost three-quarters of today's mothers in the U.S. say they played outdoors every day as children, but only one-quarter say their kids do.

SOURCE: CONTEMPORARY ISSUES IN EARLY CHILDHOOD (2004)

Give children time and space to play.
Despite our culture's preference for cramming schedules with sports, music lessons, and playdates, children thrive on unstructured play. When left to their own devices outdoors, imagination takes off. Kids who feel free to love what they love, whether it's digging in the dirt or staring at clouds, are more likely to see outside time as something to look forward to rather than a boring activity.

Listen for and support children's preferences.
Talking to children about their time outside can be as important as discussing what they learned at school. When a child feels your interest, it helps them understand that outdoor experiences matter and will help them find creative ways to celebrate it.

Empower kids to find nature anywhere.
It's true that all cities are not created equal when it comes to open space. However, having lived both on a farm and in cities, and having helped raise children in both kinds of settings, I know what most parents know—that nature is all around us if we just take the time to look.

▶ Stroll around the neighborhood, listening for birds, watching for insects, and collecting "treasures" like leaves and rocks.
▶ Encourage activities like tree climbing and stargazing.
▶ Avoid the "look, don't touch" mentality. Promote a healthy curiosity about the natural world.
▶ Plan special trips to outdoor-oriented institutions like botanic gardens, arboretums, and parks.

2. Track your family's carbon footprint and waste.

Even young children can learn the basics of a carbon footprint. You can turn individual lessons and chores into a meaningful family activity by discussing your household's larger impact on the Earth.

Frame the discussion.

Explain the big picture of how household energy use adds up within the home as well as how individual homes contribute to the collective problem of global climate change. Impress upon children the importance of individual responsibility when it comes to ethical energy consumption. Teach them the basics about the global energy system, including the importance of energy efficiency and the advantages of renewable sources of energy compared to dirty fossil fuels.

Discuss how your household currently measures up.

Review your utility bills to find your average monthly usage. Visit an online carbon calculator like the one provided by the Environmental Protection Agency to learn the carbon footprint of your home. Based on what you find, set a goal for future savings.

Map out where you use energy.

Make it a game to seek out every way your house uses electricity. Watch for standby energy devices like computers or cable boxes, which suck energy even when they're not in use. *See page 261 to learn what to do with this problem.*

▶ For information visit: youth.zerofootprint.net and climatekids.nasa.gov.

QUICK TIP

The young reader's editions of my books *An Inconvenient Truth* and *Our Choice* are also good resources.

3. Promote climate change education.

Over the past 20 years, environmental education programs have had major benefits, from improved critical thinking to increased civic engagement. Unfortunately, more often than not, climate change education is lacking. A recent survey found that 30 percent of science teachers who teach climate change tell students that it is "likely due to natural causes," while another

31 percent teach the issue as unsettled science. Parents should encourage teachers to resist the effort by climate deniers to undermine science.

Here are some suggestions you can make to your children's teachers, or use if you're a teacher yourself:

Turn to technology.
Teachers can bring science to life with engaging apps, like BBC's *Earth*, NASA's *Earth Now*, or WWF's *Together*, available on iTunes, Apple TV, Google Play, or your local cable television.

Bring in the arts.
Climate change is interdisciplinary. Art teachers can assign climate change posters. Creative writing teachers can assign poems on the subject.

Explore these climate and energy curriculum resources.
- ▶ climatekids.nasa.gov/menu/teach
- ▶ nwf.org/Eco-Schools
- ▶ eia.gov/kids/energy.cfm?page =teacher_guide ◉

Haven Coleman, a young Climate Reality Leader, asks her congressman about renewable energy policy at a town hall meeting.

Colorado Springs, Colorado
April 13, 2017

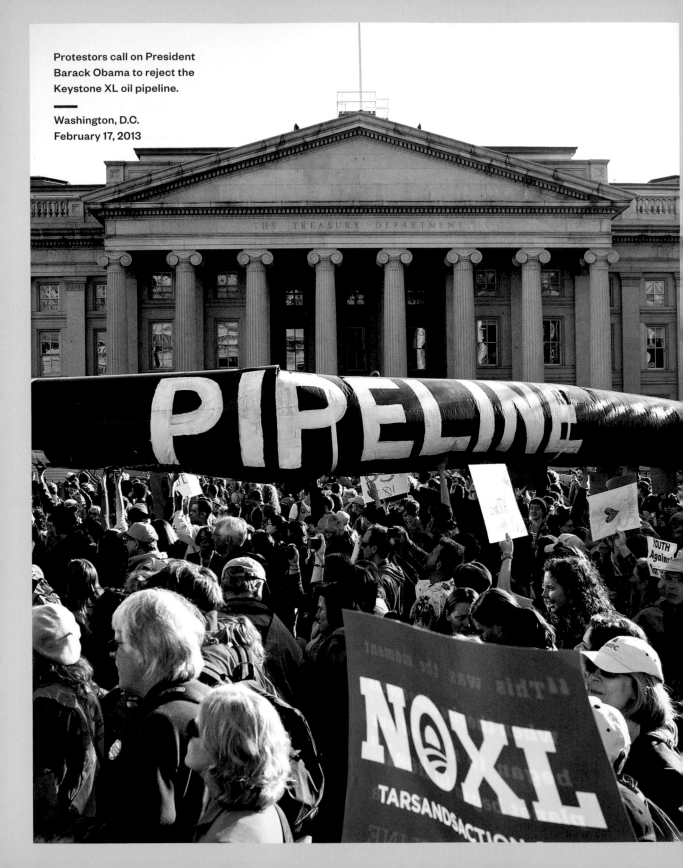

Protestors call on President Barack Obama to reject the Keystone XL oil pipeline.

Washington, D.C.
February 17, 2013

Harriet Shugarman

Founder of Climate Mama and Climate Reality Leader

WYCKOFF, NEW JERSEY

HARRIET SHUGARMAN SPENT years building a career at the United Nations and the International Monetary Fund, where her environmental work included helping organize the first Earth Summit. But it wasn't until the births of her children that the magnitude of the climate crisis washed over her in full force.

"There was so much happiness, but also such a sense of responsibility," said Shugarman, who grew up in Alberta, where oil and gas are a large part of the economy. "Learning about the climate crisis, you might want to forget it, but you know it's real."

After the release of *An Inconvenient Truth*, Shugarman attended one of the first Climate Reality trainings in Nashville. When she returned to New York to deliver her own presentations at schools and community centers, she discovered that her identity as a mother gave her a powerful ethos as a speaker.

Though she'd spent many years working with diplomats and officials at the UN and the IMF, Shugarman recalls that her climate presentations gave her a new sense of effecting direct change in the world.

"It took me a year or two after giving presentations in my community to realize that I could speak in this really personal voice as a parent," she said. "I felt empowered."

As she discovered that voice, Shugarman founded an organization called Climate Mama that publishes information about environmentalism and advocacy aimed at parents—and has been recognized by the White House and the Environmental Protection Agency for its outreach efforts.

Shugarman also keeps busy as an activist. In 2011, she was arrested in Washington, D.C., at a protest against the Keystone XL Pipeline. She also teaches about climate change policy at New Jersey's Ramapo College, writes for national outlets including MSNBC and *The Huffington Post*, and until recently chaired the Environmental Commission and Green Team in the New Jersey town where her family lives.

"I am hopeful," she said, "because of the many amazing people I meet from the climate movement and the environmental movement and beyond. I'm hopeful because of my children. I can't let myself not be hopeful." ◉

To learn more about Shugarman's work, visit ClimateMama.com.

Tactics and Strategies of the Anti-Smoking Movement

HOW CAN CLIMATE ACTIVISTS challenge and counteract the propaganda coming from billion-dollar fossil fuel industries, as well as the constant denial from "skeptics"? The campaign against tobacco use shows us how difficult the challenge can be—but also what success can look like.

In 1998, the "truth" campaign was launched, with the goal of ending teen tobacco use. The theme of the campaign was exposing tobacco industry tactics that manipulated teen consumers—tactics that we now know were also adopted by large carbon polluters in the fossil fuels industry to mislead the public about the true causes of climate change. (Fittingly, the first years of funding came from a massive settlement between 46 U.S. states and the biggest tobacco companies.)

The truth campaign marked a significant change from prior anti-smoking campaigns, which had focused on negative health effects. Extensive research had shown that teens were not dissuaded from smoking by warnings about lung cancer and other consequences. So the truth campaign used a different approach in radio, television, and billboard advertising to cleverly highlight the blatant falsehoods of the tobacco companies and provoke teens to ask themselves a subversive question: "Am I being played for a fool and taken advantage of by these giant corporations?"

Additionally, it used "horizontal influence," meaning not a top-down, but peer-to-peer approach. The campaign coordinated regional youth-led groups called SWAT, Students Working Against Tobacco. According to a 2009 issue of the American Journal of Preventative Medicine, truth prevented 450,000 youth from starting to smoke within the first four years of the campaign.

In 2014, the Centers for Disease Control and Prevention launched the "TIPS" campaign to urge adult smokers to quit. The ads showcased former smokers suffering from tobacco-related illnesses giving "tips" to current smokers for how to handle the ailments when they start suffering from them. The campaign inspired almost two million people to quit over three years.

Corporate Accountability International (CAI) found success using boycotts against tobacco companies. CAI launched a boycott against Kraft products, owned by cigarette producer Philip Morris (now Altria; the campaign was so effective that the company changed its name!). The campaign particularly focused on boycotting Kraft Macaroni and Cheese to put pressure on the tobacco industry to end marketing targeted at young people. The boycott tarnished Kraft's image as a family brand. Boycotts have been effective in the climate movement too—for example a "switch your account" campaign led Bank of America to limit its financing of mountaintop removal coal mining practices.

The fossil fuel industry has misled much of the public into thinking climate change is not real—much like the cigarette companies promised us their product wasn't dangerous. The truth campaign show that through innovative and emotional activism and marketing, people can be persuaded. This achievement provides hope for making the truth about the climate crisis just as persuasive. ◉

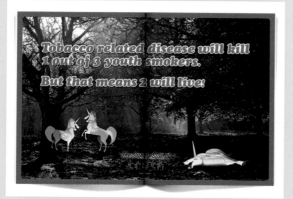

Images from "The Sunny Side of Truth" campaign, launched in 2008.

Talk with Climate Deniers

Even though climate science is peer-reviewed, comprehensive, and well established, I know all too well that it is still a deeply polarizing subject. Climate deniers and those of us who accept the facts are sometimes pitted against one another at family gatherings, social occasions, and in the larger public discourse. It's important that you not only understand why this controversy persists (hint: it's manufactured) but also that you can confidently and meaningfully address these skeptics.

It is important to remember that this divisiveness has been purposefully engineered by oil, coal, and gas companies—and the politicians whose pockets they line. In Tennessee, we have an old saying, "If you see a turtle on top of a fencepost, you can be pretty sure it didn't get there by itself." In the same way, when you hear all the climate denial in the U.S., you can be pretty sure it didn't get there by itself either. Over the past few decades, large carbon polluters have invested heavily to "manufacture doubt" and advance obfuscation instead of science—in much the

same way the tobacco lobby successfully created doubt in the public's mind as to whether or not cigarettes cause cancer. Ultimately, they lost that battle, but not before they delayed public health responses to cigarette smoking by over 40 years. The carbon-polluting lobby is using their playbook to do the same where global warming is concerned.

It's now clear that Exxon Mobil knew about the climate consequences of burning fossil fuels beginning in the late 1970s—but instead of changing course based on its own research, the company instead poured millions into efforts to dupe the public about those consequences. We also know that President Donald Trump once called climate change a Chinese hoax (although he later claimed he was joking), and the man he appointed to head the Environmental Protection Agency, Scott Pruitt, even denied the most basic scientific finding concerning global warming: that CO_2 emissions trap heat in the atmosphere. That fact was proven by scientists more than 150 years ago!

These misinformation efforts have thus far been all too effective. They've obstructed progress by normalizing outright denial of science so effectively that the U.S. has more climate change deniers than any other country.

Don't be dissuaded if some denier says, "I don't want to talk about it." Don't let that stop you.

Speak up.

And they have created a political and cultural divide so deep that the very topic of global warming has become almost taboo for some to discuss in polite company. According to a 2015 study, 74 percent of Americans rarely, if ever, discuss climate change. It is difficult to dispel misinformation about something you are not supposed to talk about. This makes it all the more important for climate activists to break the silence and engage in civil debate with those who have been misled.

The goal of this section is not to encourage you to convince every climate denier you run into; it is to make you feel confident that if a lively debate is occurring, you're armed with the facts and reasoning to set the record straight. Rest assured that the science is on your side. You just have to remember some basic questions and responses, brought to you in part by the Climate Reality Project. SkepticalScience.com is also a fantastic resource for novel and effective ways of interacting with climate deniers.

1. How can there be global warming when it's snowing outside?

We still have winter—and we still have "natural variability." This means that the fluctuations from season to season and from day to day still exist. And that means

that some days will still be cold. But if you look at the bell curves on pages 46 and 47, you will see that, on average, there has been an enormous increase in hot days and extremely hot days, while there has been a sharp decrease in the number of cold days. In fact, some say that's one of the reasons we notice the cold days a lot more when they do show up.

Global warming is about broader, long-term climate trends scientists are seeing around the world, as opposed to local weather patterns. Daily temperatures may rise and fall, but it's the long-term heating trend that matters. There are multiple signals of this long-term trend: winter comes later and spring comes earlier. Summers are longer and hotter. All of the natural systems of the Earth—including plants and animals—are forced to adjust to the dramatic changes underway. All of the ice-covered regions of the world are melting and sea level is rising. It is getting so hot in some regions of North Africa and the Middle East that scientists are now predicting that parts of the Earth that are now heavily populated will become uninhabitable.

2. Why do some deniers claim that the Earth has been cooling since 1998?

To begin with, all the temperature records around the world demonstrate very clearly that this assertion is completely false. It's true that 1998 was a warm year, and for a few years after 1998, its temperature record was not broken. However, the record set in 1998 has since been broken many times. The hottest year of all was 2016; the second hottest was the year before; the third hottest was the year before that. You can find the graph on pages 50–51. If the person claiming the world is cooling also tells you that the moon landing was faked, it might be time to end the conversation. But do

it respectfully and politely—just don't waste any more time.

3. How do you explain the fact that not all scientists agree the climate is changing?

Multiple peer-reviewed studies confirm that 97 percent of the world's top climate scientists publishing articles in peer-reviewed journals agree that man-made pollution is the principal cause of global

DENIERS CAN MISLEAD BY CHERRY-PICKING SHORT TIME FRAMES

- Realist's view of global warming
- Denier's view of global warming

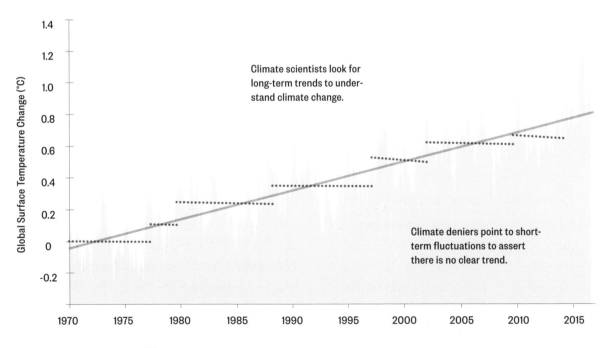

Climate scientists look for long-term trends to understand climate change.

Climate deniers point to short-term fluctuations to assert there is no clear trend.

SOURCE: SKEPTICALSCIENCE.COM

warming. In fact, during the past few years, less than 0.1 percent of climate research rejects the consensus. Every national academy of science in the world agrees, as does every major scientific organization. Several of the tiny group of scientists who disagree with the consensus have shadowy financial ties to the fossil fuel industry.

Here, I find an analogy works well. For example, if you have chest pains and were somehow able to get advice from the 100 leading heart doctors in the world, what would you do if 97 of the 100 expressed alarm and told you that you needed to go to the hospital right away? Would you do nothing because three of the 100 doctors say they just aren't sure and you don't need to do anything?

4. Natural cycles have always influenced planetary climate. How can you really believe humans could possibly have that much impact?

Humanity has become the largest force of nature affecting the ecological system of the Earth because of three factors working together: First, we have quadrupled our population in less than a century; second, we now have at our disposal technologies far more powerful than any that people in prior eras could

have dreamed of; and third, our obsession with short-term thinking in the new hyper-global economy has served to blind us to the impact of what we are doing—most of all in using the most vulnerable part of the Earth's system, the atmosphere, as an open sewer.

While it's true that natural cycles have an impact on climate, human activities are now overwhelming all of those natural cycles—primarily because we are spewing 110 million tons of man-made global warming pollution into the atmosphere every 24 hours and it is trapping as much extra heat energy as 400,000 Hiroshima-class atomic bombs exploding every day. But again, man-made global warming pollution now overwhelms all the natural factors put together.

5. Plants need carbon dioxide, and we need plants. So how is more CO_2 supposed to be a bad thing?

Carbon dioxide is necessary for life on Earth. We all need it to help retain the sun's heat, and plants of course need it to "breathe." Some plants, it's true, do appreciate an extra dose of carbon dioxide. But others are damaged by excess CO_2, which reduces the level of nutrients in many food crops and increases the damage done by many plant pests.

A person looks out at Glacier Blomstrandbreen in the high Arctic.

Svalbard, Norway
1922

A Greenpeace campaigner looks
out at the now retreated Glacier
Blomstrandbreen in the high Arctic.

Svalbard, Norway
July 2002

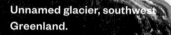

Unnamed glacier, southwest Greenland.

—

**Ujaraannaq Valley, Greenland
Summer 1935**

4

Same location, southwest Greenland.

—

**Ujaraannaq Valley, Greenland
Summer 2013**

Moreover, even those trees and other plants that can theoretically grow faster with more CO_2 cannot do so without concomitant increases in soil nutrients and other factors also necessary for a higher growth rate.

In addition, the disruption of the water cycle by excessive CO_2 levels in the atmosphere hurts many food crops. Already, we're seeing major devastation to many valuable food crops, including reduced yields from corn and wheat due to the heat stress directly linked to excessive CO_2.

6. We all breathe out carbon dioxide. Are you saying we should all stop doing that, too?

Let's all take a breath here—because breathing has nothing to do with global warming.

Each of us breathes out about 2.2 pounds of carbon dioxide a day, which might sound like a problem when you consider there are more than 7 billion of us doing it all day, every day.

But that's not the case, because we are participating in a closed cycle. That is, the carbon we breathe out comes from the food we eat, whether directly from plants, which take it in via photosynthesis, or indirectly from meat. So we get carbon from plants, breathe it out, and they take it in again.

Burning dirty energy is a completely different story. When we dig up and burn coal, oil, or gas, we are bringing carbon into the air that has been stored underground for millions of years. And once we've released that carbon genie from the bottle and into the atmosphere, it can remain there for hundreds of years—some of it for thousands of years.

7. How can a couple of degrees be such a big deal?

The average human body temperature, at 37°C (98.6°F), is much higher than the current average global surface temperature on Earth—14.8°C (58.7°F). The 20th century average was 13.9°C (57°F), and yet an increase of 2°C (3.6°F) in your body temperature is a reason to go to the doctor and see what's wrong. And if it's not a temporary increase but a long-lasting one, and if it continues to increase day after day, the doctor is likely to tell you that you have a serious health problem. The same is true for the Earth. We have given it a fever, and the fever keeps going up. All of the delicately balanced ecological systems are stressed by the increasing temperatures.

A scientist releases a weather balloon, which measures atmospheric pressure, temperature, and humidity.

Antarctica

Climatologists predict that a 2°C (3.6°F) rise above the pre-industrial average would have devastating effects on plants and animals, food crops, and public health. And unfortunately, that is exactly where we are heading in just another 30 years. Unless we stop loading the atmosphere with even more heat-trapping pollution, temperatures could increase by 4°C to 5°C (7.2°F to 9°F) in the lifetime of the millennial generation.

Also, bear in mind that the problem is not just a problem of surface temperature—it is also about the many other consequences of a 2°C (3.6°F) increase, such as the disruption of the water cycle, the accelerated melting of ice leading to sea

235

level rise, and all the other consequences described in Part I of this book. What's more, a quick review of historical climate trends shows us that even the increase of less than 1ºC (1.8ºF) since 1980 has already caused a dramatic increase in extreme weather events, from intense rainstorms and severe droughts to more frequent heat waves.

These events will only worsen unless we act now.

8. Won't limiting emissions also limit jobs and economic growth?

A fast growing category of new jobs is in renewable energy, efficiency improvements, and the Sustainability Revolution. For example, solar jobs in the United States are growing 17 times faster than average job growth, and the fastest-growing job occupation is wind turbine technician. Numerous economic studies have now proven that limiting emissions does not limit economic dynamism. The good news is that exactly the opposite is true.

And the economic losses of climate change are mounting, already costing the world as much as $1.2 trillion per year. Every year, climate change is responsible

for 400,000 lives lost, from mega-storms like Superstorm Sandy and Typhoon Haiyan to the impact of vector-borne diseases—not to mention the millions of deaths from conventional air pollution generated by the burning of fossil fuels.

Maintaining our dependence on dirty fossil fuels would not only devastate public health and the environment, it would also prevent us from gaining the many economic benefits of the job-intensive transformation to a clean energy economy.

One study, for example, found that decarbonizing the electric system would boost global GDP by $19 trillion between now and 2050.

9. If warming is inevitable, even if we went 100 percent renewable, then what is the point?

Yes, climate change is happening. We have already suffered some regrettable losses, and unfortunately, we cannot go back in time to stop what we have started. However, we can still prevent the catastrophic damage that we would otherwise suffer by acting boldly and quickly to limit and then reduce green-house gas emissions.

"Obstructionist attitudes, even on the part of believers, can range from denial of the problem to indifference, nonchalant resignation or blind confidence in technical solutions. We require a new and universal solidarity."

—Pope Francis

10. For the sake of argument, let's say I get on board with climate science. What can one person do?

The good news is that there is plenty we can do now to help reduce the impact of climate change.

Each of us can take steps individually, but all of us must act as citizens to ensure that our government—and others around the world—takes collective action to reduce our dependence on fossil fuels and move to a clean energy economy. This means that we not only have to change our light bulbs, we have to change our laws. We can also choose the products we buy with an eye to minimizing the impact on the climate. This sends a powerful signal to business and industry that they need to offer more climate-friendly goods and services.

For more information
- ▶ Check out the "Truth in 10" slideshow at inconvenientsequel.com
- ▶ Look for the Skeptical Science app on iTunes or Google Play for on-the-go, science-based responses to skeptics.
- ▶ You can back up your arguments with graphs and charts available from NASA at climate.nasa.gov/earth-apps.
- ▶ The Climate Reality Project has answers for questions from deniers at climaterealityproject.org/content/12-questions-every-climate-activist-hears-and-what-say. ◎

Oren Lyons Jr.

Faithkeeper of the Turtle Clan of the Onondaga Nation of the Iroquois Confederacy

ONONDAGA, NEW YORK

OREN LYONS JR. WAS THE ELDEST of seven siblings who grew up on an Onondaga reservation in upstate New York. In the early 1940s, when he was a young teen and his father left the family, it became his responsibility to provide for the family by hunting, fishing, and gathering firewood.

"I was pretty much hunting for the family," said Lyons, who is now 87 years old. "Our life was tough, hard, but we had a lot of fun. We got to a certain age where we were in the woods all day long."

From early on, it was clear that Lyons was a talented artist. After serving as a heavy machine gunner in the 82nd Airborne Division during the 1950s, he was accepted to Syracuse University on a lacrosse scholarship, where he studied at the College of Fine Arts. He was an all-American, on the same team with the legendary Jim Brown. Lacrosse, of course, was invented by the Iroquois. When he graduated, he moved to New York City and stayed at the YMCA, determined to become a professional artist.

He eventually found a job as a commercial artist for Norcross Greeting Cards, where he rose through the ranks to become the director of art and planning. He married and bought a house in New Jersey.

Everything changed when his aunt, who served as a Clan Mother back home, asked if he could take a leadership role in the Turtle Clan of the Onondaga Nation of the Iroquois Confederacy. Lyons was taken aback; in many ways, he'd traded that life for the American Dream. He took a year to research the history, culture, and cosmology of his ancestors, then decided to accept a position on the Council of Chiefs.

In the end, the combined workload of his corporate job and Council responsibilities was too much. "I thought I could stay in New York and do it, but it didn't work out," he said. "It was too much, and I had to go home."

In the wake of that decision, he embraced his role as an advocate for the Iroquois. He became a professor of Native American history and culture at the University of Buffalo, protested development projects on indigenous land, and attracted the attention of John Lennon and Yoko Ono, both of whom visited Lyons in Onondaga in 1971.

Lyons also helped found the Traditional Circle of Indian Elders and Youth and traveled internationally to learn about issues facing indigenous people around the world. He wrote multiple books, edited a magazine called *Daybreak*, and in 1977 joined a delegation in Geneva that successfully lobbied the United Nations to recognize the status of indigenous people.

Over the years, Lyons's attention turned to the environment, and to the connections between

Lyons participates in the People's Climate March.

—

New York, New York
September 21, 2014

climate change and the rights of native people.

"I think the point of no return for global warming is much closer than people think," he said. "I've been up to Greenland twice now. You want to get scared, take a look at what's going on up there."

Lyons points to the Onondaga's yearly cycle of ceremonies—when the sap begins to flow from the maple tree, when planting begins, at the first harvest, at the midwinter—as the type of touchstone that has kept the Iroquois society in harmony with the Earth for thousands of years.

Native people have suffered terribly during centuries of colonialism, he argues, and the effects of environmental degradation are likely to be the most severe for disenfranchised populations. Lyons wonders whether the world's policy makers might benefit from the traditional wisdom of indigenous

groups. He reflects sometimes on an ancient Iroquois law that suggests considering the effects of current actions on the people who will be alive in seven generations.

"When I was growing up as a kid, seven generations were easy to see," he said, sighing. "It was like forever. Now, it's pretty god darn murky."

Lyons takes strength, he says, in his memories of his close-to-the-Earth childhood on the Onondaga reservation and in the rich tribal traditions he rediscovered over the course of his life.

"We were instructed as leaders: never take hope from the people," he said. "So regardless, you've got to strive on." ◉

An employee checks a cable
on a Vestas wind turbine.

Osterild, Denmark
April 18, 2016

Find a Career in Renewable Energy

You can dedicate your life to fighting climate change without becoming a full-time volunteer or sacrificing your livelihood. For example, solar jobs in the United States are growing 17 times faster than average job growth, and the fastest-growing occupation is wind turbine technician. Whether you're in corporate headquarters or out on the solar or wind farm, a job in renewable energy is a great way to fight climate change every day.

There are two main ways to break into this exciting industry:

1. Use your technical talents.

Are you an electrical contractor and want to diversify your skills? If so, there are exciting new opportunities to adapt your technical abilities to the clean energy revolution. Many have found it valuable to become certified as a solar technician.

▶ The U.S. Department of Energy lists places across the country where people can get various types of training for renewable energy jobs at energy.gov/eere/education/find-trainings.

▶ If you wish to pursue certification, solar installers should visit nabcep.org/certification and wind installers can explore energy.gov/eere/wind/wind-testing-and-certification.

2. Use your other skills.

Renewable energy companies have many needs beyond technical expertise—from sales and marketing to human resources. This means there are many opportunities to work for firms helping to solve the climate crisis even if you do not have technical skills.

▶ Non-technical types of renewable energy careers include:
  – Accounting and administration
  – Research and data analysis
  – Procurement and outsourcing
  – Construction
  – Sales and marketing
  – Support services from facilities management to legal support and recruitment

QUICK TIP

Get inspired by real-world stories of people joining the solar economy at *www.stories.solar*, a project of the Southern Environmental Law Center.

Once you've chosen a direction, make the switch:

Network.
Attend renewable energy events and conferences to educate yourself about the industry and meet new people. Learn from and introduce yourself to the experts; show them you are curious and motivated to participate.

Follow industry news.
Read industry publications and follow leaders on social media. This will help you to show depth of knowledge in your cover letters and interviews.

Go to school.
Consider taking an elective in sustainable business at your local college or through an online course.

Be an activist.
Your skills may speak for themselves, but your passion for advancing climate solutions can help set you apart from other candidates in your field. Practice explaining why you want to work in this industry. Stand out as a well-informed, well-intentioned activist who is as excited about fighting climate change as you are about being productive in the business side of companies that are part of the solution. ◉

A solar technician at work.

We are going to win this; we are going to solve this crisis.

The key is solving it quickly enough. That's why it matters that so many people are organizing to accelerate the solutions.

Wei-Tai Kwok

COO of Amber Kinetics, Inc. and Climate Reality Leader

LAFAYETTE, CALIFORNIA

WEI-TAI KWOK SHOULD HAVE been happy. His family life, with a wife and two school-age children, was fulfilling. He had built his Bay Area firm, Dae Advertising, into an industry power player with a client list that included Disney and Apple.

But when he saw *An Inconvenient Truth*, it set him on a course that would come to change everything. In the days after the screening, he found himself preoccupied by the ways global climate change would affect the lives of his kids, then aged nine and six. As he tends to do, he confronted his concerns by researching them meticulously—and what he found angered him. He was perturbed not just at the policy makers who, he came to believe, weren't doing enough to cut carbon emissions, but also at himself for a lifetime of complacency.

"I got mad," he said. "I realized that not only was I not part of the solution, but I was part of the problem. It became an ethical issue to me that was intolerable. I could not enjoy being at work anymore."

Kwok attended my 2013 Climate Reality Leadership training in Chicago, where he learned to apply his experience in corporate communications to giving presentations about climate science. Afterward, he set the goal of reaching 1,000 people in his hometown of Lafayette, Louisiana, and when he met it, he doubled his goal—and reached it—the following year.

Kwok also decided to dedicate his professional life to climate progress by leaving his advertising firm for a series of jobs in the clean energy sector, including stints at Suntech Power, NRG Energy, and Andalay Solar. He's currently serving as the chief operating officer for Amber Kinetics, a Union City, California, startup working on a mechanical battery that stores energy using a flywheel.

Now, working with renewable energy during the day and as a Climate Leader after hours, Kwok has turned the climate crisis into an opportunity to build a brighter future for the Earth. He credits that growth to the diversity and camaraderie of his fellow climate leaders in Chicago.

"It was so inspiring," he said, "to sit with people of all ages and genders and creeds and work together to solve this problem. It filled me with hope." ◉

To learn more about becoming a Climate Reality Leader, see page 297.

Make Your Business More Sustainable

Businesses have the opportunity to be on the front lines of the Sustainability Revolution. Whether you are a business owner who can make the decisions yourself or a dedicated employee willing to push for change from within, every business can and should pollute less and help fight the climate crisis by achieving maximum efficiency and leveraging the passions of its employees to make it happen.

1. Make the business case.

From those with entry-level jobs all the way up to those in the C-suites, every employee of a company can encourage more sustainability. In most cases, success depends upon winning support from colleagues, managers, and other stakeholders. Here are some arguments that can help you explain how sustainability efforts benefit the business where you work:

Sustainability can reduce operating costs and improve business intelligence.
By using smart building technology to facilitate and monitor energy use, operations managers can not only ensure that systems like heating and cooling run more efficiently, they can also use predictive analytics to drive future performance and planning.

Attracting consumers and employees.
In a recent poll, 64 percent of global CEOs said that Corporate Social Responsibility goals were core to their business, in part because they help drive public and partner trust. Furthermore, brands with demonstrated sustainability goals are seeing a growth in consumer sales. That means promoting a climate agenda can improve your reputation and support profitability at the same time. This is also important for attracting and retaining employees who want their work to reflect their values, an increasingly important factor for job satisfaction among younger employees.

Sustainability is more affordable than you think.
Renewables and "smart building" technology—everything from lighting sensors to solar systems—are getting

Solar panels mounted above a parking area serve double-duty providing shade for the cars below.

A new cycle bridge called "The Bicycle Snake" provides increased accessibility and safety for cyclists.

Copenhagen, Denmark
July 14, 2014

less expensive every year. Plus, there are many valuable legal and financial incentives for energy efficiency programs, such as tax credits, fee reductions, waivers, and even expedited building permits.

Turn wasted space into energy and value.

An empty rooftop or vacant parking lot sizzling in the sun can be covered with solar panels.

▶ More details for building your case can be found at usgbc.org/articles/business-case-green-building.

Position yourself as an early leader.

The need for businesses to embrace a sustainable approach is growing due to increasing external forces, including policy changes like a price on carbon, pressure from large investors to disclose climate risk (especially if their supply chains are vulnerable) and to identify opportunities for emissions reductions, and rapid technological changes that are undermining existing business models. This is why major corporations worldwide are moving quickly to adopt renewable energy and to innovate with new products that address fast-growing markets focused on sustainability.

2. Invest in clean energy and efficiency.

Buildings use about 40 percent of the total energy consumed in the U.S., making them a vital place to focus our efforts. Here are some ways to improve energy use and reduce your organization's carbon footprint:

Assess the situation.

Start by understanding where you are now. Benchmark your energy use by identifying the ratio of economic output to emissions or researching the ratio of electricity consumption to employees in your company.

▶ A great resource for per-employee benchmarks is g20-energy-efficiency.enerdata.net/indicators/unit-electricity-consumption-of-services-per-employee.html.

▶ You can also use a carbon footprint calculator to assess your baseline; try the one at coolclimate.berkeley.edu/business-calculator.

Identify areas of opportunity.

Approximately 30 percent of energy used in buildings is wasted, so there are likely many ways you could improve efficiency. Here are a few places to look:

The building envelope.
Leaky or poorly insulated walls, windows, and doors waste energy. Updating these thermal barriers can keep the building operating at higher efficiency. An EnergyStar roof can reduce peak cooling demand by 10 to 15 percent.

Building systems.
Heating, cooling, and ventilation (HVAC) comprises about 34 percent of an average commercial building's energy use. EnergyStar systems typically use 10 to 20 percent less energy than conventional products. Lighting and computers also require a lot of energy, so look for savings potential here, too, from ensuring the use of LED bulbs to investing in "smart lighting."

Energy supply.
If onsite development of renewables is not an option, look at your utility's green power options. Your firm may also be able to purchase renewable energy credits.

▶ Check for potential incentives in your state through the Database of State Incentives for Renewables and Efficiency at dsireusa.org.

Build your strategy.
Set clear goals for energy savings, tactics, and timelines. Identify project stakeholders, communicate the plan clearly, and work together—for example, to research HVAC systems that would work for your office. Decide who will be responsible for advancing the plan and how to measure progress. When you reach your goal, celebrate!

3. Make it easy to be green.

While energy systems for buildings are a huge part of the equation, day-to-day use also adds up. Try these ideas to make your workplace more conducive to climate-friendly behavior:

Empower recycling.
Make sure your workplace has easily accessible recycling bins.

Choose green supplies.
Make sure the supply closet is stocked with environmentally friendly options, such as refillable toner and inkjet cartridges, reusable dishes, and Forest Stewardship Council–certified paper. Buy in bulk to save on packaging.

Reduce paper waste.
Encourage colleagues to reduce paper use in simple ways, such as printing double-sided and using online collaboration tools that reduce the need for paper documents.

Prevent e-waste.
Most electronics include harmful materials that can be recycled safely.

"What is the business case for an economic system that says it is cheaper to destroy the Earth than to take care of it? How did such a fantasy system that defies common sense even come to be? How did we—all of us—get swept up in its siren's song?"

—Ray Anderson, late chairman of Interface, Inc.

▶ The EPA lists major nationwide retailers that offer computer, battery, and cell phone recycling options at epa.gov/recycle/electronics-donation-and-recycling.

Engage with your colleagues.
Survey people to find out what actions they think are important, then use that knowledge to design a program that best reflects your organization's values. Ongoing idea exchanges via internal meetings as well as lunch-and-learn seminars with experts will keep your colleagues connected.

Create a climate action team.
Form a group that brings together the most passionate people from across the organization, ideally including at least one person from each department. Define your purpose, outline key goals and tactics, and designate roles.

Organize actions and initiatives.
Rally people around special events, such as Bike to Work day. Look for creative ways to incentivize participation.

Promote climate-friendly benefits.
Among the employee benefits that can also help the planet are pretax public transportation passes, tuition support, climate-related performance bonuses, and socially responsible investment choices within a pension or 401(k) program. ◉

Run for Office

You might be surprised to learn how many opportunities there are to become an elected government official. While there are only 537 federal offices in the U.S., there are more than half a million opportunities to run for office in the more than 90,000 government units, including city and county positions, school boards, water districts, and so on.

You don't need a degree in political science or a track record in local politics to get elected. In fact, the House of Representatives is made up of an increasingly large share of people from business backgrounds, and a broad mix of professions can be found in any public office, from English teachers to scientists. Each office comes with its own opportunity to influence.

1. Assess your prospects.

Visit your city, county, or state Board of Elections to learn which offices will be on the ballot soon. When choosing the office you'd like to run for, consider whether factors like your skills, experience, and network could resonate with voters for one particular office over another. It's also worth considering whether you are

A woman waves a flag at a presidential rally at the Santa Monica High School Football Field.

Santa Monica, California
May 23, 2016

interested in challenging an incumbent, or running for an open office. Then think about your answers to some relevant questions, such as: Can you meet the eligibility requirements? Are you ready to look people in the eye and ask them for their support?

Although running for office has in recent years required a lot of fundraising, this appears to be changing. Some candidates for office during the last election—including one prominent candidate for U.S. president— demonstrated that it is now possible

"Progress occurs when courageous, skillful leaders seize the opportunity to change things for the better."

—Harry S. Truman, Former U.S. President

Baltimore mayoral candidate DeRay Mckesson canvasses in the Charles Village neighborhood.

Baltimore, Maryland
March 26, 2016

to raise money on the Internet in small amounts from lots of people. In addition, campaigns for local offices typically cost much less. Many candidates have been elected on shoestring budgets. Make sure that you have support from your family and others who could be personally affected by your decision to run for office.

2. Lay the groundwork.

Get organized well in advance of making any formal announcement of your candidacy. It is critical that you become visibly active in the community, attending important events, meeting with influencers, and talking with people about the issues you and they care about. If you are new to town, or have been otherwise engaged, it is never too late to get involved. Act now, enlisting your family and close friends to help you in these early days.

3. Build your campaign team.

Campaign teams vary in size depending on the campaign. The key roles you may need to fill include a campaign manager, directors in charge of finance, communications, politics, and outreach, as well as a volunteer coordinator and a treasurer.

4. Craft your message.

Why should you win? What are your goals while in office? How will you see them through? These are questions you and your campaign will have to answer every day, in every encounter. Create a simple, memorable, and unified campaign message to ensure voters know who you are and what you stand for.

To develop that message, you need a strong sense of who your target voters are, and how to appeal to them. Study voter demographics in your area. Who do you need to win over?

5. Rally support.

Once you have a campaign strategy, it is time to get the word out there and connect with as many voters as possible.

Do this by prioritizing:

▶ Door-to-door canvassing and phone banking
▶ Online engagement
▶ Town halls and other public speaking events
▶ Fundraisers
▶ Coalitions and endorsements
▶ Advertising and PR

Running for office is hard work—as is being a public official. In terms of making direct and lasting impacts in the fight against climate change, though, there's hardly a better choice to make.

One more thing—and it's actually the most important thing—always remember why you got involved in your campaign in the first place and hold true to your values. Don't ever compromise your integrity. Stay focused on your objectives to bring positive change. ◉

Talking with voters during my first U.S. Congressional campaign.

—

Rural Middle Tennessee
July 29, 1976

Steven Miles

Minister for Environment and Heritage
Protection, Minister for National Parks and the
Great Barrier Reef, and a Climate Reality Leader

QUEENSLAND, AUSTRALIA

ABOUT 10 YEARS AGO, when Steven Miles's wife was pregnant with their first child, he started to worry about climate change. The couple intended to raise a family in Australia's coastal Queensland, and the threat of rising sea levels began to alarm him.

"Climate change is already affecting Queensland," Miles said. "We're having hotter and hotter days, more cyclones. My motivation in politics was always about future generations—like education and the economy—but when you think about it, all of those things will be impacted if we can't stop the planet from getting hotter."

After Miles saw *An Inconvenient Truth*, he traveled to Melbourne to take part in the Climate Reality Training program I led there in 2007. Afterward, while working day jobs for trade unions and nonprofits and completing a PhD in political science, he became engrossed with Queensland's lush coastal biosphere—and how it has come to be menaced by the changing climate.

He became especially concerned with the Great Barrier Reef, a marvel of the natural world that runs down the length of Queensland's shoreline. Scientists believe it to be the largest structure on Earth made of living organisms, and its rich ecosystem shelters many endangered and vulnerable species of whales and sea turtles.

The reef is also, Miles learned, straining under unprecedented environmental stress. The warming ocean has caused a series of catastrophic "bleaching events," which occur when higher temperatures cause the disruption of a symbiotic relationship between the coral and beneficial organisms called zooxanthellae, which provide nutrients to the coral. If the damage continues unchecked, the entire ecosystem could collapse. The bleaching started in 1998 and continues to this day; in the summer of 2016, the reef suffered its most serious bleaching event in history.

"It was catastrophic," Miles said. "The amount of coral that hasn't recovered is enormous."

By 2014, Miles and his wife had three children, and it was on their behalf that he mounted a successful campaign for parliament under the slogan "Miles Better for the Reef." His former colleagues were supportive of his environmental message, with one union leader from his days in labor advocacy quipping that "there ain't no jobs on a dead planet."

When he was elected in early 2015, Miles was appointed environmental minister, and he immediately dedicated himself to protecting Queensland's natural habitats. He increased spending on koala conservation, pushed for mining and land rehabilitation reform, and, of course, worked diligently to protect the Great Barrier Reef.

TRUTH TO POWER

Bleached coral from a mass-bleaching event in 2017. Pixie Reef, near Cairns Australia, March 3, 2017

Queensland alone can't change the global climate, but Miles hopes that if the state can reduce strain on the reef through local legislation that fights fertilizer runoff and other threats, it will buy the reef enough time for countries around the world to unite against carbon emissions and stop or reverse the coral bleaching for good.

"We're actually putting a lot of stress on the reef through localized pollution," he said. "The scientists tell me that if we can take the pressure off from that, we can give the reef extra time to confront global warming."

With the reef's future teetering on the brink, he also hopes that Queensland can reduce its own carbon footprint. Australia is largely dependent on fossil fuels for electricity and transportation, but Miles believes the country is poised for an energy revolution as it invests in a mix of photovoltaic, solar-thermal, and wind infrastructure for a greener future.

In the service of those goals, Miles maintains a frenetic schedule. When we spoke on the phone for this book, he had to take a short break in the middle for a live interview with a local radio station about water-quality testing in the Fitzroy River Basin. Miles sees the Paris Agreement, which Australia ratified late last year, as a watershed moment in the international struggle for climate progress. He senses a momentum that he's never felt before, especially among business leaders, who, he believes, are starting to see environmental responsibility as a corporate duty.

"I've felt a real shift post-Paris in the business sector," he said. "Multinational companies want to be in business in 2050, so they have long-term plans, and they're building into their plans the targets that were set in Paris."

Miles's advice for climate activists around the world who want to make a change in their home communities is simple: start educating yourself on how to run for office, even if it's just a local position, in order to push for positive change at a grassroots level.

"I am optimistic," he said. "I think you have to be." ◉

Walk the Walk

While it's great to encourage your family, friends, and community to be more climate conscious, it's important you, yourself, practice what you preach. You should lead by example by shrinking your carbon footprint and making climate-friendlier consumer choices. People will expect it, and they should. Furthermore, it will help you understand the intricacies of sustainable living and better inform your decisions.

1. Evaluate your current impact and set a target.

Calculate how much energy you use by looking at your utility bills and setting a specific goal for reducing it: for example, 10 percent in six months.

▶ It might help to use the EPA's Carbon Footprint Calculator at epa.gov /carbon-footprint-calculator.

2. Audit your home.

Search your home for opportunities to reach your goal. Inspect each room, making a list of everything that is plugged in. Look for and note any outright energy drains, such as drafts.

Consider bringing in a professional to conduct an even more thorough evaluation, including inspecting major appliances and identifying areas that are missing insulation. The Department of Energy estimates that following a professional auditor's recommendations for efficiency upgrades can result in saving up to 30 percent on your energy bill.

▶ Learn more at energy.gov/energysaver /professional-home-energy-audits.

3. Look for ways to save.

Heating, cooling, and hot water.
Keeping a home heated and cooled comprises about half of the energy use in a typical American house, while water heating represents about 18 percent. There are simple ways to lower those figures. Use a programmable thermostat and make sure your water heater is insulated against heat loss. You may want to go further and invest in new, more efficient systems, which can often save more money than the purchase price in just a few years. In addition, solar water heaters are eligible for tax credits through 2021.

Any place with an outlet.
The average household owns dozens of consumer electronics, which add up to 12 percent of a home's electricity use.

▶ Consider updating to more efficient options, which you can find at energystar.gov.

QUICK TIP

Use power strips to cut the flow of electricity to standby devices, like TVs and computers, which can suck up energy even when they are off.

It is estimated that if every TV sold in the U.S. was EnergyStar–certified, 9 billion pounds of greenhouse gas emissions would be averted each year.

Garden.
Planting the right trees around your home will help clean the air—and can add valuable shade that can reduce energy bills. Tree cover can naturally trim summer air-conditioning bills by up to 35 percent. Consider starting a compost pile, which helps save the energy used to transport food waste and keeps decaying organic matter from contributing methane emissions to a landfill.

Garage or driveway.
First, leave the car behind as often as you can. There are also several ways to improve your carbon footprint when

Preschool students plant apple trees in an event organized by the German Forest Protection Community for Day of the Forest.

Near Baerenthoren, Germany
March 21, 2017

using your vehicle, such as avoiding idling and keeping your tires inflated to the right pressure, which can save up to 10 percent on fuel. When it is time to replace your current vehicle, choose a more efficient option, ideally an electric vehicle.

▶ Research available federal tax credits for EVs at fueleconomy.gov /feg/taxevb.shtml.

4. Make the switch: consider going solar at home.

You should start by asking your electric utility to switch to clean energy. Google has a great tool called Project Sunroof which shows you your solar options.

▶ To learn more, visit google.com/get/sunroof.

A tractor pulling a weed harrow
on farmland.

June 19, 2012

Eat with the Planet in Mind

The EPA has reported that all the emissions from the "electric power, transportation, industrial, and agricultural sectors associated with growing, processing, transporting, and disposing of food" account for a significant percentage of U.S. emissions. Whether it's by eating less meat or buying local, you should consider agriculture and its pollutant by-products whenever shopping or eating out.

1. Eat less meat, especially red meat.

Eating a lot of meat results in more greenhouse gas (GHG) emissions due to the inefficient transfer of plant energy to animal energy, which creates a larger impact on the climate per calorie. Cutting down on the amount of meat you eat, especially red meat, can have a significant effect on emissions—and, doctors say, also improve your health.

2. Make plant-based food a bigger part of your diet.

Because of meat's heavy GHG emissions, lowering your intake even slightly can have an impact. Going one or two days a week without meat is a good start. Plant-based foods like lentils, beans, and whole grains are great sources of protein. If you do want to eat meat, try swapping out beef and pork for chicken, dairy, eggs, or fish, which have much lower carbon footprints.

3. Go local.

Transportation makes up of about 11 percent of food-related emissions. Eating only local foods for a year would avert the carbon emissions associated with driving 1,000 miles in a passenger vehicle.

▶ Use localharvest.org to find nearby sources of local foods, including farmers' markets.

4. Buy foods that don't contribute to deforestation.

Deforestation contributes 10 percent of all GHG emissions. Check the websites of the brands you use to be sure they have pledged to deforestation-free supply chains, and check Greenpeace's scorecards of how they are really doing.

5. Buy in bulk.

After you've found your favorite low-emission food items, consider buying more of them. Buying more food at one time, assuming you eat it all, will save you trips to the grocery store, and, in turn, reduce your emissions.

6. Buy organic.

A study found that organic agriculture uses 30 to 50 percent less energy than traditional agriculture. ◉

The Boqueria market.

Barcelona, Spain
March 1, 2014

"Food production is a main driver of biodiversity loss and a large contributor to climate change and pollution, so our food choices matter."

—Bojana Bajzelj, Researcher at University of Cambridge

Morning at the Rodale Institute,
an organic farm since 1947.

—

Kutztown, Pennsylvania
October 18, 2016

Support Organic Agriculture

FARMERS STAND ON THE FRONT LINES of the climate struggle—not only because they often feel the brunt of severe weather and drought, but also because they have the power to help reverse climate change with every crop they grow.

Since humans began farming the earth, the world's soils have lost 30 percent to 75 percent of their carbon content. Not only does this mean that some of that carbon is now in the atmosphere, contributing to the climate crisis, but it also means less fertile soil and lower yields. Depletion of soil carbon has worsened with intensive, modern farming practices like the dependence on synthetic nitrogen fertilizer, tillage, and monocropping, all of which disrupt soil's natural carbon storage capabilities.

The good news is that the soil offers a truly down-to-earth opportunity to fight climate change. According to the Rodale Institute, an agricultural research nonprofit based in Kutztown, Pennsylvania, by using organic, regenerative soil management methods, farmers can play a significant role in taking carbon out of the atmosphere and putting it in the ground, converting it from a greenhouse gas into a food-producing asset. The organization's research shows that by switching current croplands and pastures to organic management practices, we could potentially sequester a nontrivial percentage of current annual CO_2 emissions.

Regenerative organic agriculture comprises a range of time-tested practices that contribute to soil's ability to retain carbon. Many of these are easy and relatively inexpensive to adopt. For example, cover cropping, mulching, and composting all help keep organic matter (i.e., carbon) in the soil. They also include conservation tillage or no-till farming, which can play a vital role in keeping plant residue in the soil and reducing erosion. Regenerative, organic agriculture can be economical too because it helps farmers maintain yields and improve farm profitability.

Meanwhile, demand for organic produce has been rising dramatically over the past decade, with pesticide-free fare now sold in most mainstream grocery stores. According to the Organic Trade Association, in 2015, Americans spent more than $43 billion on organic products, and nearly 22,000 businesses earned organic certification, up a record 12 percent from the year before.

Still, the organic market is dwarfed by conventional food production which currently accounts for more than three quarters of all fruit and vegetable sales. So it's important that we use our hard-won dollars to help show the broader industry that organic is not just a nice-to-have—it's better for the Earth and, for many consumers, it's a must-have.

Restructuring our global food system is a massive undertaking, and one that each of us can contribute to by voting with our forks and our wallets. Educate yourself on where your food comes from by reading labels, getting online to learn more, and talking with your local farmers about their climate-friendly practices. ◉

Learn more at rodaleinstitute.org/regenerative-organic-agriculture-and-climate-change.

When you insist on buying the most climate-friendly goods and services, you add to the pressure on businesses to continue leading the Sustainability Revolution.

A customer browses produce in a local grocery store.

—

Yangon, Myanmar
March 11, 2017

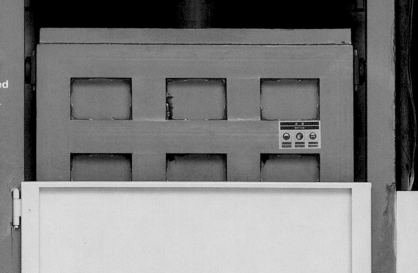

Cotton scraps collected by the TAL Garment Manufacturing group and remade into reclaimed cotton clothing by Patagonia.

Vote with Your Dollars

The industries that profit from the destruction of our planet rely on your hard-earned money to justify their fossil fuel habits. One of the most basic and potent forms of power we have is our ability to decide where and how we spend our money. Regardless of your financial situation, there are some simple ways to make sure you're supporting businesses that are part of the solution rather than the problem.

1. Study before you buy.

You already research the products and brands you buy for price. You can also examine their business practices. Look for companies that have made an explicit and authentic commitment to sustainability and eco-friendly practices.

2. Invest with climate in mind.

Investment is a powerful tool to support eco-friendly companies. And if you have investments in polluting companies,

Yvon Chouinard, founder of Patagonia, has proven that corporations can be sustainable as well as profitable.

———

Ventura, California
January 23, 2010

divesting can be a strong form of protest, as well as a wise move. Over the past few years, large investment fund companies have lost billions due to their investments in fossil fuels. Seek out financial advisers and investment firms that have made a strong commitment to eco-friendly values. Look closely at your investment portfolio to make sure all the companies you invest in follow meaningful "Environmental, Social, Governance" (ESG) criteria. Companies that meet rigorous ESG criteria are more likely to be socially responsible environmental stewards and have an ethical corporate culture.

3. Make more efficient travel plans.

Travel is an area where almost every family can easily reduce its carbon footprint.

▶ Look for hotels that have green business practices, such as those who participate in the Green Hotels Association, which you can learn about at greenhotels.com.

Some airlines are better than others, or offer carbon offset options. Try to fly less overall. Consider exploring national forests or state parks near your home-town for vacation.

4. Green your banking.

Keep your money with institutions
that support climate advocacy work.
Some organizations, including CREDO
Action, the Sierra Club, and the League
of Conservation Voters, offer their own
credit cards, so your interest payments
go to support environmental activism
instead of Wall Street. Smaller banks
and credit unions are more likely to fund
clean energy and local advocacy work.

5. Give to the movement.

Reducing your own environmental
impact is just a start. We all need to
support the groups and nonprofits that
spend every day fighting for the future
of our planet. Set personal goals for how
much you can donate every month. Pick
organizations that are doing work you
believe in, and where your money will
make the greatest difference. ◉

How to Create and Host Events

The author Malcolm Gladwell wrote an influential essay a few years ago called "Small Change," which perceptively described the relative weakness of any strategy for organizing that puts its sole emphasis on clicking boxes on the Internet. Online organizing plays a crucial role, but in order to develop a stronger sense of cohesion and commitment among those you are organizing, it is essential to hold in-person events. Marches, concerts, film screenings, town hall and city council meetings, teach-ins, and student rallies can have enormous impacts—including recruiting new activists, generating media coverage, and putting pressure on decision makers.

The Live Earth concert, which inspired people to find and act on solutions to climate change.

Rio De Janiero, Brazil
July 7, 2007

1. Have a Strategy.

Getting 100, 500, or 1,000 people to come to a rally or other event is great, but if you don't have a plan to leverage that attendance into political power and action, it might be a waste of time. Before you start organizing, take the time to identify some clear goals, such as recruiting 5-10 new members or volunteers, collecting 100 postcards to send to legislators, raising a certain amount of money, or gathering 50 new email addresses or phone numbers.

Think about how achieving these short-term goals will help you achieve your longer-term goals, like getting a policy changed or dramatically increasing public awareness of a climate issue.

2. Choose what kind of event you'd like to put on.

Rallies, marches, and demonstrations.
Be sure to set reasonable goals. Think of how you can make the event visually

Demonstrators march on the British parliament as part of the Time to Act march.

London, England
March 7, 2015

compelling and successful even if only a small number of people show up. You don't need large numbers of people to get attention for your cause. Get creative and consider using costumes, props, and theatrics to convey your message. Most importantly, make sure that your demonstration is seen and heard by the people you're trying to influence and inspire. You can mobilize on the sidewalk outside a corporate headquarters or on the steps of your city hall. If possible, organize the demonstration during business hours so that the decision makers you are trying to influence will see it. If you choose a weekend to increase attendance, be sure to alert the media and collect photos and video to send to your target and the press and to share on social media. The Climate Reality Project also hosts regular Days of Action—try planning an event on one of these days to increase attention by linking to a larger organization.

In the U.S., the First Amendment protects the right to assemble. In most cities, you can organize demonstrations on public sidewalks without any need for a permit. Larger demonstrations and marches that take over the street often do require a permit. Most city government websites have some information about permits for public events, so do your research to avoid any conflict.

Boycotts.
If a business in your community is engaging in irresponsible practices that are harmful to the planet, you may want to take the extraordinary step of organizing a boycott.

However, once you call for a boycott, you need to back it up. Organize a picket line outside the business you are boycotting and have volunteers pass out leaflets encouraging customers to spend their money elsewhere. Make sure to have clear, achievable demands so that the company knows what they need to do to get the boycott to stop. Your goal is to make it more economically or politically costly for them to continue the negative practice than it is for them to agree to your demands.

Letter-writing parties and phone banking.
As noted on pages 186–189, it's important to regularly contact your elected officials and other decision makers, but it's a lot more fun—and a lot more effective—to organize an event with a group of people who are doing the same. Provide materials for people to use to write letters or send postcards to their representatives, or printed scripts for people who want to make phone calls. Serve refreshments, and keep it light and fun.

Women grab postcards during a local postcard-writing event.

Boulder, Colorado
January 29, 2017

"We cannot condemn our children, and their children, to a future that is beyond their capacity to repair."

—Barack Obama, Former U.S. President

Help others learn more about the climate crisis and its solutions.
Show your support for climate action by going to see *An Inconvenient Sequel: Truth to Power* at your local theater. Bring your family, friends, and peers to share the experience and start a conversation about how your community can get involved.

After *An Inconvenient Sequel: Truth to Power* is released for home viewing, host a movie party with your friends or family. Be sure to open the floor to discussion after the film, and have a facilitator or expert there who can keep the conversation productive and offer concrete action steps. If you would like to host a public screening of either of my films, contact Paramount Pictures at (323) 956-5000 and ask for the Repertory/Non-Theatrical Department.

Climate Reality presentations.
The Climate Reality Project gives presentations focused on climate science and climate action. At a Climate Reality Leader presentation, participants learn about the impacts of climate change on a local and global level, solutions to solving the climate crisis, and what you and your organization can do to fight climate change and create a better future for the planet.

▶ You can request a free Climate Reality presentation by going to realityhub. climaterealityproject.org/events /attendpresentation. Once you've requested a presentation, follow the event hosting advice above to promote it.

Benefit parties.
You probably already have organizing experience, even if you have only hosted a small gathering of friends. A benefit party isn't much different—you just ask your guests to make a donation to the cause. You can have someone collect donations at the door or "pass the hat." Offer refreshments, put on music, and tell your friends to invite their friends. These events can be casual, but take some time to talk to the guests one-on-one and as a group. Explain what you're raising money for, why it's important, and how they can get more involved.

QUICK TIP

See page 297 to learn how to become a Climate Reality Leader and make your own climate presentations.

Fundraising meals.

Consider organizing a breakfast, lunch, or dinner event to raise funds. Cook a tasty and inexpensive meal yourself, or ask local restaurants to donate or offer discounted food to support the cause. Charge a flat rate or sliding scale donation for entry to the meal, and give your guests opportunities to contribute more if they can. Some restaurants will let you host fundraising events in their space, and donate a portion of their profits.

Concerts and performances.

Partner with local or national artists who are sympathetic to your cause and who can draw a crowd. They may be willing to donate their labor, but you will often want to offer them a percentage of the money raised or an honorarium. If you treat the artists well, they will be more likely to promote your event to their fan base. It's often helpful to partner with a local venue or a promoter who has experience with local venues, sound equipment, and local music press.

Just because there's music doesn't mean your event can't also include short, upbeat speeches or announcements from local groups between musical acts. Have a charismatic host to remind the audience of the purpose of the event and give them ways to take action or get involved. If you're partnering with a well-known artist or band, contact local media outlets and offer them an interview with your performer before or during the event. Prep the performers with talking points ahead of time.

3. Get people to come.

Every event is different, and you'll need to make decisions based on your goals, your audience, and your resources. But there are some key "best practices" that apply to almost any event, whether it's a loud demonstration or a black-tie gala. Every organizer who has hosted events has had the fear, "What if no one comes?" To avoid this scenario, it's important to recognize how much work it takes to turn large numbers of people out to an event. You'll need to use every tool in the toolbox, whether you're trying to fill a venue that holds 50 people or 5,000.

Here are some basics for every event:

Don't go it alone.

Your events will be more successful if you have help organizing them. Find collaborators who can bring their own ideas and social networks to the table. Partner with other organizations and divide up the responsibilities related to the event.

Pick the right venue.

Where your event happens is important.

Continued on p. 286.

Indivisible

Leah Greenberg and Ezra Levin, Co-Founders

WASHINGTON, D.C.

LEAH GREENBERG AND EZRA LEVIN are two of the former congressional staffers behind the Indivisible Guide, an online handbook and website dedicated to empowering activists to resist the Trump administration's agenda. It's an effective group: in fact, my youngest daughter, Sarah Maiani, is a member of the Indivisible group in Santa Barbara, California. Earlier this year, Greenberg and Levin, a married couple, sat down with me to talk about the origins of Indivisible and the most effective ways you can engage with your local representatives.

AL GORE: You've cited the Tea Party as an inspiration for Indivisible. Why do you think the Tea Party was so effective, and what can other activists learn from it?

EZRA LEVIN: After the election, we realized that, putting aside the Tea Party's racism and violence, they were smart on strategy. They implemented a defensive strategy focused on their own representatives. They went to public events and town halls and made calls. It worked. They took down a lot of the agenda of Barack Obama, a historically popular president with an enormous congressional majority.

LEAH GREENBERG: When I was working for Representative Tom Perriello (D-Virginia), the Tea Party took up a lot of mental space in the office. Whatever we were doing, you had to think about what their response was going to be.

AG: What are some examples of effective activism you've seen since the 2016 election?
EL: One thing that blew us away was the first congressional recess, in February 2017, when every member of Congress goes back home to listen to their constituents. Local Indivisible groups have been doing phenomenal work in helping people organize and attend town halls. In response, many Republican members have actually refused to hold town halls. We've even seen people holding their own town halls: if their representative showed up, great; otherwise, they were ready with a cardboard cutout or an empty chair.

LG: The thing that has struck me is the power of stories. In early January, a group of constituents brought into Virginia Representative Barbara Comstock's office a family that was threatened by the repeal of the Affordable Care Act. This family had a 10-year-old daughter who has a preexisting condition; without treatment, she would die. These stories can change what is seen as possible.

Indivisible volunteers hold hands at the Women's March.

Washington, D.C.
January 21, 2017

AG: What's the story behind the name "Indivisible"?

LG: We were trying to think of something that was rooted in American history, but also encompassed the full diversity of America today. The only way to respond to a demagogue who is threatening components of our civic society is to treat an attack on one as an attack on all.

AG: How can the way a congressional office responds to an outcry by its constituents have an impact on the way elected officials make decisions?

EL: Members of Congress want to be reelected. That means they care about how their constituents view them and how the local press treats them. Our theory of change here is that any individual member of Congress cares much more about their image back home than about any one thing that President Trump is trying to get done.

AG: What steps can activists take to avoid burnout?

EL: That's front and center in our minds right now. The scenario we want to avoid is that two years from now, the story becomes, "Remember back in 2017, when we all did activism for a while?" These are everyday citizens who are giving up their nights and weekends to fight for things they believe in. It's important to build a sense of community.

AG: If someone reading this book wants to become politically active, what should they do?

EL: Go and get involved in your local Indivisible group. Go to IndivisibleGuide.com and type in your zip code. No matter where you are in the country, you will find a group near you. Get involved on your own turf. ◉

Remember the old rule of thumb: make sure you can fill the space. A large crowd overflowing a small space sends a powerful message. The same size crowd, dwarfed by a large venue with lots of empty seats sends a weak message. Pick a central location that's easily accessible by public transportation and/ or has ample parking available. Spaces that are already well known to the activist community, like progressive churches (or other houses of worship), bookstores, music venues, or community centers, are often best. College campuses may have a variety of spaces that you can rent or reserve for free through a student organization or friendly professor.

Promote shamelessly.
You need to be a champion of your event. Create a Facebook event and invite all your connections. Then message the guests and ask them to invite their friends. Post the link to your

During the COP21 conference, demonstrators paint the roads surrounding the Arc de Triomphe with non-polluting water-based paint to create the image of a sun.

Paris, France
December 11, 2015

event in relevant online groups and forums, email it to your mailing list, and ask local organizations to share it with their members. You can also create an online form using a service like Action Network, NationBuilder, or Google Forms where people can RSVP for your event by providing their name and contact information. That way you can email them the day before the event to remind them or send them relevant information. For larger events like mass demonstrations, you may want to set up a custom webpage with information about the event and an easy way for people to sign up for updates. Hand out flyers or postcards at other similar happenings, on college campuses, or at the grocery store.

Leverage word of mouth.
Directly invite as many people as possible, and get commitments from them that they'll be there. Telephone

"We are the first generation that can put an end to poverty and we are the last generation that can put an end to climate change."

—Ban Ki-Moon, Former Secretary General of the U.N.

trees also work extremely well. Asking someone to bring something, like snacks, or to play a volunteer role at the event is a great way to make them feel more connected—and committed.

Utilize advertising and media.
If you have a budget, even a small one, consider Facebook or Twitter ads to promote your event to a targeted audience, or consider taking out an ad in your local paper or on your local radio station. Often, local TV stations will mention your event as a public service, if you let them know about it in advance. Then move on to getting media coverage. Most local outlets have event listings that you can submit to for free. Contact local journalists who cover politics and the environment and encourage them to write about your event in advance, or make it one of their "weekly picks." See page 211 for more tips on getting press coverage.

Create compelling visuals.
Make your event stand out. Take the time to create attractive posters, handbills, and online graphics. You don't need to be a designer—free online services like Canva.com have templates that make this easy for anyone (though it won't hurt to enlist a designer on your team). Don't crowd your outreach materials with information—just provide the key details, and direct people to an email address or website where they can learn more. See pages 290–291 for Shepard Fairey's advice for designing fantastic demonstration materials.

3. Manage a successful event.

Regardless of size, the day of the event will probably be a little stressful and require some running around to take care of logistics. Make sure you have a good team of volunteers to help, and that everyone knows what they are responsible for ahead of time. If a few people show up to your event early, put them to work. Here are some roles you will likely need to fill:

▶ Greeters to welcome people as they come in, ask them to sign in with their name, email, and phone number, and collect donations if applicable.

▶ Hosts and emcees to make announcements, keep the program running smoothly, and keep the crowd energized and engaged.

▶ Servers and bartenders to help sell or give away food or drinks.

▶ Canvassers to work the crowd asking people to take various types of actions like signing a postcard to local lawmakers, opting in to receive text messages, or signing a paper petition. This may not be necessary at smaller events, but it is key for larger and outdoor events.

▶ Security might be needed if you're concerned about counter-demonstrators or someone disrupting your event. It's always good to have a few people who are on the lookout for potential problems, and they should be trained in de-escalation and conflict management to help quickly address anything that comes up.

▶ Stage managers to make sure that speakers, performers, and others are where they need to be when you need them there, and to keep things running smoothly.

5. Design great event materials.

Put your message front and center at every event. Be sure to have signs, banners, stickers, buttons, or T-shirts available for your audience to show their support. You can also provide posterboard and markers for people to make their own signs.

Here are some basic tips for designing effective materials:

▶ Keep it simple: you want your message to be immediately understandable and legible from a distance, so keep it short, to the point, and bold.

▶ Use large fonts and contrast to make your words stand out. Black letters on a white background, or vice versa, are always a good way to go.

▶ Get creative: humor, compelling images, and memorable slogans will help you reach more people. Spend some time brainstorming ideas with others.

▶ Don't break the bank: you can get banners, signs, and other materials professionally printed, but it's expensive. Consider hosting a banner-painting or sign-making party a week before your demonstration or event. Not only will you save money, you'll build excitement for the event itself. ◉

Shepard Fairey

Artist and Founder of OBEY Clothing

LOS ANGELES, CALIFORNIA

SHEPARD FAIREY IS A PIONEERING figure in the street art movement and an outspoken advocate for social and environmental justice. His work has achieved near ubiquity both through his company, OBEY, and the now famous 2008 "Hope" poster depicting Barack Obama. Early in 2017, we sat down to discuss ways that artists and activists can work together to fight climate change.

AL GORE: How did you first become concerned with the climate movement?
SHEPARD FAIREY: I had an abstract understanding of how our growing population, on a planet with finite resources, was a problem. But really, it was seeing *An Inconvenient Truth*. It had a dramatic impact on me. The entire presentation was very persuasive, and it led me to do more research, and it woke me up to the urgency of addressing climate change.

AG: Your work has raised awareness of the most pressing social issues of our time. Based on your experience, what is the role of art in inspiring social activism?
SF: I think the great thing about art is that it affects people emotionally, so it can break through predispositions. If a verbal or written argument doesn't

register, it's easy to block it out. When art connects to something human, that's where the magic is.

Shepard Fairey painting "Liberté, Égalité, Fraternité" mural in Paris, 2016.

Fairey paints a Project C:Change Mural called "Peace Elephant."

Hong Kong
October 25, 2016

AG: What advice would you give to climate activists who want to incorporate art and design into their activism?

SF: One of the challenges with climate change is that it's a big issue with a lot of different variables. It's hard to nail down just one symbol for it, and people often try to get too many ideas into one piece. I'd encourage people to convey something relatable. Let a smokestack be a symbol. Let rising sea levels be a symbol.

AG: You've used the word "propaganda" to describe your work. What do you mean by that? And to turn our last question around, how can a person create effective propaganda?

SF: I try to be very transparent about what I'm doing. I use the word "propaganda" to encourage people

to be realistic about analyzing everything that comes at them, including what I'm sending their way. When a corporation or politician is saying that the science isn't solid enough on climate change, you need to look at what their real reason for that is.

AG: If you're using art to build momentum behind a movement, what's more important: quantity or quality?

SF: I have a whole section in my biggest monograph, *Supply & Demand*, called "Repetition Works." What's fascinating about virally transmitting images is that people subconsciously understand which way the tide is moving. I think pop cultural symbols, like the peace sign, had a lot to do with the move away from supporting the Vietnam War. I want to see the same thing happen with climate change. ◉

an inconvenient sequel
TRUTH TO POWER

PEOPLE'S CLIMATE MARCH

TO CHANGE EVERYTHING WE NEED EVERYONE

MARCH TO SAVE THE WORLD NYC 9/21
Columbus Circle @ 11:30AM   PeoplesClimateMarch.org

TOP: A participant at the Global Climate March in Berlin, on the eve of the opening of the U.N. climate summit in Paris.

BOTTOM: Shepard Fairey's poster for the 2014 People's Climate March.

LEFT PAGE TOP ROW (RIGHT): A movie poster for *An Inconvenient Sequel*.

LEFT PAGE TOP ROW (LEFT): A San Diego rally against climate change on February 21, 2017.

LEFT PAGE MIDDLE ROW (RIGHT): A protester displays a sign at the 2017 San Diego rally.

LEFT PAGE MIDDLE ROW (LEFT): In opposition to the policies of Donald Trump, Greenpeace protesters hang a banner reading "Resist" from a crane behind the White House.

LEFT PAGE BOTTOM ROW (RIGHT): A 2015 climate march in Berlin.

LEFT PAGE BOTTOM ROW (LEFT): Over 1,000 activists protest the Dakota Access Pipeline at the U.S. District Court in Washington, D.C.

It is wrong to pollute
this Earth and destroy
the climate balance.

It is right to give hope to the future generation.

Climate Reality Leadership
Corps training.

—

Shenzhen, China
June 2016

Become a Climate Reality Leader

The Climate Reality Leadership Corps is a diverse, global network of passionate people dedicated to fighting climate change through informative presentations, events, policies, and action.

Each of the more than 12,000 Climate Reality Leaders have been through my trainings—multiday events that present not only the science of global warming, but also teach leaders essential skills in communicating the challenges and solutions to the crisis. By giving their own presentations to organizations around the world, they are building a 21st-century movement for climate action.

The Climate Reality Project's mission is to catalyze a global solution to the climate crisis by making urgent action a necessity across every level of society. One of the main ways it does this is by training a network of cultural leaders, organizers, scientists, storytellers, and other citizens around the world to explain the climate crisis and its various solutions. This book provides a glimpse of the training that Climate Reality Leaders get at two- and three-day workshops—but there is so much more to experience and learn at these events.

In the Climate Reality Leadership Corps, exceptional leaders are trained through a suite of in-depth programming on science, communications, and organizing.

1. What does it mean to be a Climate Reality Leader?

A Climate Reality Leader is a changemaker who is committed to promoting understanding of the climate crisis and its solutions—and stimulating action by those who want to help solve it. Each one has been trained by a set of experts to speak truth to power, to stand up to anti-science climate denial and the well-funded dirty energy lobby, and to motivate their friends, colleagues, and neighbors to help solve the crisis. As of this writing there are more than 12,000 Climate Reality Leaders making a difference through action and engagement with their local networks. They are inspiring communities around the world to take climate action now.

By the way, 100 percent of the profits I would have otherwise received from *An Inconvenient Truth* and the book of the same name went to finance the founding of the Climate Reality Project. The same is true of the profits from this book and the movie of the same name.

One of the principal ways Climate Leaders serve is by giving their own version of the updated slideshow about the impacts of—and solutions to—the climate crisis. They also inform their audiences about what they can do to support the shift to clean, renewable energy. We need more people involved in this fight, and spreading this information through free, engaging, targeted presentations is a crucial action.

2. How can I attend a Climate Reality Leader's presentation?

Customized Climate Reality presentations can be organized anywhere, from a library or city hall to a local influencer's living room. When you attend one of these regularly updated, multimedia presentations, you will learn about the global climate crisis, the clean

Give us three days, we'll give you the tools to change the world.

—The Climate Reality Leadership Corps motto

energy revolution, and local issues and opportunities.

▶ To learn about presentations near you, visit Climate Reality at realityhub .climaterealityproject.org/events /attendpresentation.

Here, you can sort by Title, Venue, or Description, or explore an interactive map of upcoming presentations. If you find an event you would like to attend, take a second to register for a free account, then RSVP to the presentation in your area.

3. How can I train to be a Climate Reality Leader?

Are you ready to help lead the movement? Attend a multiday Climate Reality Leadership Corps training session to learn the ropes from renowned climate scientists and communicators. I personally attend all three days of these training programs as well. Since 2006, 35 of these inspiring, hands-on training events have been held in cities from Istanbul to

Shenzhen, Jakarta to Rio de Janeiro— and of course, in cities throughout the United States.

While each training is unique, depending on when and where it takes place, they all offer in-depth learning about what is happening to our planet, as well as the real-world storytelling skills it takes to inspire others to take action.

4. What will I learn?

Several speakers and panelists are lined up for each training, from leading climate scientists who will offer the latest in data and predictive analytics, to communications experts who will train you in the skills of persuasive speaking, in-person outreach, digital activism, and media relations. A mix of hands-on workshopping, expert lectures, and meaningful networking prepare you to build broad grassroots support for your endeavors.

▶ To learn when the next training will be taking place, visit climaterealityproject.org/training.

5. What happens when the training is over?

Within a year of completing the training, each Climate Leader is expected to perform 10 "Acts of Leadership." Of course, some Climate Leaders end up performing more than 100 acts of leadership each year. These acts can take a variety of forms. For example, giving a presentation, writing a blog or letter to the editor, organizing a film screening, meeting with local, state, or federal leaders, or taking part in a day of action. Some speaking events may be arranged for you through the Climate Reality Project, but for the most part, Climate Leaders are responsible for seeking out opportunities.

Leaders can keep in touch with one another by using the Reality Hub, an online community of Climate Reality Leaders and other changemakers.

▶ Visit realityhub.climaterealityproject.org to explore and connect with a global community of climate activists. ◉

Attendees of The Climate Reality Project's 24th Climate Reality Leadership Corps training.

—

Johannesburg, South Africa
March 12, 2014

Give Your Own Climate Change Presentation

The most important task for Climate Reality Leaders is to spread the word: to give free and engaging climate change presentations to their own audiences. Much of the Leaders' training is focused on these presentations.

In conjunction with the release of this book and the film *An Inconvenient Sequel*, I am making available a compact, 10-minute version of the slideshow for you to download and use in your own climate presentations. The fight against the climate crisis needs more people—including you.

1. Get the presentation.

▶ Download the "Truth in 10" slideshow at InconvenientSequel.com.

2. Find your audience.

Your book club, your church, your family, your company—you probably already have several natural audiences for a presentation. Starting small is a good strategy, but don't be afraid to bring the presentation to bigger groups; they may be even more interested.

3. Be prepared.

This book and *An Inconvenient Sequel* are both great training for giving the presentation. Be sure to rehearse before you present. Find the right pace and try narrating it out loud, by yourself. Think about the questions you may get and be prepared with answers.

4. Make it personal.

A good presentation tells a story. While the facts of the climate crisis are compelling, your own connection to the issue and your suggestions for actions you and the audience can take are what people will remember most. Add personal observations where you can to make it clear that you have a stake in the fight. Explain (and show) why solving the climate crisis means so much to you, and how its impacts and solutions will affect your audience.

5. Manage your time, hope, and complexity budgets.

You'll have three budgets with an audience, make sure you use them carefully.

Time.
People will give you only a certain amount of their time. The more personal and interesting your story is though, the more time they will allow.

Hope.
You never want your audience to lose hope. Yet in telling the climate change story, some truths are hard to hear. Before showing examples of the suffering climate-related events are causing, I warn my audience that these examples are hard to see, but to hold on—hope is coming.

Complexity.
Everyone has a threshold for processing complex information. Tailor the overall complexity of your message to your audience, and be sure to alternate difficult concepts with information that's easier to absorb.

6. Use an array of visuals.

There are a variety of ways to make the data and your story more powerful, entertaining, and informative. Here are some examples of the types of slides I include in every presentation, and why I use them. You may want to adopt these practices when making your own slideshow.

Set the stage.
I always start my slideshows with this photo. It was the first full-disk view of Earth we ever saw, and it helped to spark a new era of environmental protections. This view reminds us that we are all connected, and all in this fight together.

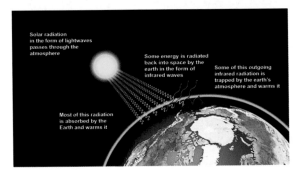

Explain the science.
Infographics like this can help to explain complicated systems or concepts clearly. Step-by-step illustrations can help to engage audience members who might tune out a complex message delivered only through text or pictures.

Fort McMurray, Alberta, Canada
May 3, 2016

Stay within the "hope budget."
Photos of recent extreme weather- or climate-related events remind people of the urgency of the crisis and the need to solve it quickly. But be judicious in the number of examples you give. It's easy to deplete your audience's "hope budget."

Isolate startling facts.
Important points sometimes deserve to stand on their own. At other times, pairing them with a relevant image can convey a deeper meaning. Here the text is enhanced with a photo of an endangered species (the golden poison frog).

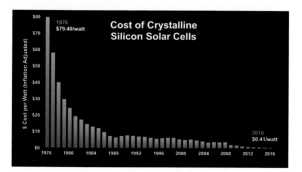

Show the proof.
Choose your charts wisely, using relatively simple examples that illustrate a dramatic change. Going overboard with charts and graphs can quickly exhaust the "complexity budget."

Use appropriate imagery.
Combine text with images that reflect a similar tone. Here, the photo gives a sense of place while the open sky and dramatic lighting signify hope. The moving boat conveys the idea that we are all moving forward, together.

7. Do it again.

Once isn't enough. You'll get better and more comfortable each time you give the presentation. It's important that we spread the truth about the climate crisis as widely as we can. The more you do it, the larger your network and influence will grow, and the more effective you will be in the fight to protect our planet. ◉

The fight against climate change will not be easy. We will encounter a series of NO's.

"After the final no
there comes a yes
and on that yes
the future world depends."

Those of us who are privileged to be alive in these early decades of the 21st century are called upon to make decisions of great consequence. Indeed, it is not an overstatement to say that the entire future of humanity depends upon whether or not we rise to the challenge before us.

The climate crisis is the most serious and threatening manifestation of an underlying collision between human civilization as it is presently organized and the ecological system of the Earth— upon which the fortunes and future prospects of our civilization, and our species, depend.

Our population has quadrupled in less than a century and is predicted to continue growing in the present century from 7.4 billion in 2017 to 9.7 billion in the next 33 years, and to 11 billion or more by 2100. Population growth is slowly stabilizing as girls are educated, women are empowered, fertility management is made widely available and child mortality continues to decline. This aspect of our relationship to the Earth is, in spite of the great challenges growing populations will pose in some regions, a success story unfolding in slow motion.

But the impact we have on the natural systems of the Earth is magnified enormously by the awesome power of the technologies that have become available to us since the Industrial and Scientific Revolutions. In particular, any decision to continue relying on dirty and polluting carbon fuels threatens to massively disrupt the climate conditions that have given rise to the flourishing of civilization and have supported the rich and diverse web of life that is integral to our survival. Global warming is the most threatening part of our ecological crisis because the thin shell of atmosphere surrounding our planet is the most vulnerable part of the Earth's system.

But there is a third factor that has led to this crisis, one that is more consequential than either population or technology. It is our way of thinking and the values on which we base the decisions we make. In particular, short-term decision-making is now commonplace in politics, culture, business, and industry. And it is now abundantly clear that

A march to protest the Dakota Access Pipeline.

——

Standing Rock Sioux Reservation, North Dakota
September 9, 2016

if we continue to ignore the long-range consequences of our present actions and behaviors, we will put our future at dire risk.

The good news—the exciting news—is that we already know that we can change the way we think. We know it with certainty because we have made historic changes in our ways of thinking before. Every great moral cause in human history was initially launched at a time when the overwhelming majority of men and women believed that the change called for was not only impractical but completely implausible. As the late Nelson Mandela said, "It is always impossible until it is done."

This aphorism is true for the abolition movement, the women's suffrage movement, the civil rights movement in the United States, the anti-apartheid movement in South Africa, and more recently, the gay rights movement in the United States and in nations around the world. As each of these moral causes gained more supporters, the changes they called for were met with increasingly fierce opposition. Each renewed

call to do the right thing was met with a resounding "No!"

And in every one of these historic struggles, those fighting for justice, faced with a seemingly endless and implacable resistance, came to doubt that victory would ever come. In the bleakest hours of the U.S. civil rights struggle, Martin Luther King Jr. answered some of his followers who plaintively asked how long it would be before they won.

"How long?" he replied, "Not long! Because no lie can live forever.... How long? Not long! Because the arc of the

moral universe is long but it bends towards justice. How long? Not long!"

A mere five years ago, if someone had predicted that in the year 2017 gay marriage would be legal throughout the United States and would be not only supported but honored and celebrated by two-thirds of the American people, I would have responded by saying, "I hope so, but I'm afraid that is extremely wishful thinking."

The pattern is always the same: once the underbrush of obfuscation, straw men, and distractions are cleared away and the underlying issue is

TRUTH TO POWER

resolved into a binary choice between what is clearly right and what is clearly wrong, then the outcome becomes preordained—because of who we are as human beings. And then the change comes quickly. As the late economist Rudi Dornbusch once observed, "Things take longer to happen than you think they will, and then they happen faster than you thought they could."

We are close—very close—to a similar tipping point in the great moral cause that is the climate movement. Every day now, millions more are awakening to the realization that it is wrong to destroy the future of the human race, and it is right to give future generations the well-being, justice, prosperity, and hope to which they are rightfully entitled.

It is also important to note the relationship between solutions to the climate crisis and the current state of the global economy. There is, at present, a growing concern about the weakness of "secular demand" throughout the global economy. The recovery from the Great Recession, which began in early 2009, has not created enough new jobs to boost incomes—and spending—in the U.S. and in many other countries. As a result, many economists have expressed concern that the global economy is in danger of slipping into another recession. Moreover, the stagnation of wages since the mid-1970s is believed by most to be a principal underlying cause of the political unrest fueling the rise of populist authoritarianism.

Meanwhile, in addition to the impact of hyper-globalization, the accelerating impact of intelligent automation is continuing to exert downward pressure on wages and is continuing to eliminate jobs in a pattern that convinces many observers that conventional economic theory—which tells us that automation always creates more jobs than it eliminates—may no longer be valid. And the reason seems to be that the extension of cognitive capacities along with physical capacities is a game changer.

In these unusual and new economic circumstances, what is most needed to restore strength to the global economy and restore confidence in the efficacy of self-governance is a coordinated global initiative to create tens of millions of new jobs throughout the world—jobs that are not vulnerable to either outsourcing or intelligent automation.

As luck would have it, the steps necessary to solve the climate crisis are exactly the same steps that would save democracy and economic prosperity. They include: a coordinated effort to retrofit buildings in communities throughout the world; an acceleration of the transition to renewable sources of energy and higher levels of efficiency in industry and business; and a shift to sustainable transportation, agriculture, and forestry. An initiative including these steps would simultaneously heal the climate crisis and become the smartest global economic strategy we could follow.

The generation of young people who will fill these new jobs are even now joining this struggle and bringing fresh resolve—reminding us of the special role that young people have so often played in focusing the attention of their elders on the clear distinction between right and wrong.

I vividly remember when I was 13 years old, hearing President John F. Kennedy commit the United States to the inspiring goal of putting a man on the moon within 10 years. And I remember how many of my elders in 1961 felt that goal was unrealistic and perhaps even impossible. But eight years and two months later, Neil Armstrong put his foot on the surface of the moon. Two seconds later, when the news of that history-making step reached NASA's mission control center in Houston, Texas, a great cheer went up—and the average age of the systems engineers cheering in that room was 26—which means that when they heard President Kennedy's challenge, they were 18 years old.

They changed their lives to gain the skills to match their inspiration and become a part of history. And many of today's 18-year-olds are doing the same. Many years from now, when they reach the age of their parents today, they will inherit the Earth we bequeath to them. And depending on the circumstances in which they find themselves, they will ask one of two questions.

If they live in a world of stronger storms, worsening floods, deeper droughts, mega-fires, tropical diseases spreading throughout vulnerable populations in all parts of the Earth, melting ice caps flooding coastal cities, unsurvivable heat extremes in the tropics and subtropics, hundreds of millions of climate refugees generating political disruptions and threatening the collapse of governance—if they face these horrors and the others of which scientists are now warning, they would be justified in looking back at us and asking,

Hundreds of thousands take part in the People's Climate March.

—

New York, New York
September 21, 2014

"What were you thinking? How could you have done this to us?"

But if they live in a world filled with a sense of renewal, with hundreds of millions of new jobs created in the Sustainability Revolution, with cleaner air and water and the growing prospect of restoring the climate balance—if they have hope in their hearts and can experience the joy of telling their own children that their lives will be better still—then they will ask a different question of us: "How did you find the moral courage to change, boldly and quickly, and save our future?"

The time for us to answer that question is now—by seeking the truth about the reality we are confronting, by using the power we all have to bring about the necessary and urgent changes, and by never forgetting that the will to change is itself a renewable resource. ◉

It's time to speak truth to power.

These are some of the 12,000 Climate Leaders giving the presentation around the world.

Join the fight at
InconvenientSequel.com.

ACKNOWLEDGMENTS

There are a great many people without whose help I could not have written this book. First of all, Liz Keadle has helped me throughout the entire process. I am also grateful to each of my children, Karenna Gore, Kristin Gore, Sarah Gore Maiani, and Albert Gore III, for advice and counsel.

I benefitted from the invaluable research by my personal staff, particularly Brad Hall and Alex Lamballe, without whom this book would never have been possible. Jill Martin worked tirelessly with me on the design of the graphs, pictures, and other graphic material. The entire staff in my Nashville office, led by my superb Chief of Staff, Beth Prichard Geer, helped enormously. And thanks to Joby Gaudet, who has worked extremely hard to ensure the constant flow of information between me and my team. Special thanks as well to Betsy McManus for all of her help and advice.

The climate scientists and other experts who have done the research on which much of this book is based are too numerous to list by name here, though many of their names are in the credits at the bottom of graphs and quotations from studies throughout the book. I would like to single out Jim Hansen, whose scientific work has been especially compelling to me for many years. I would also like to once again acknowledge the seminal work of the late Roger Revelle, who first enlightened and inspired me in the late 1960s and whose path-finding research, along with that of

Charles David Keeling, is considered by many to be the baseline for modern climate science.

This book was brilliantly edited, produced, and art directed by Melcher Media. I have worked with the CEO and founder, Charles Melcher, for many years. This is our third book together, and each project has deepened our friendship. It has been a joy to work with Charlie and his skillful team, including especially Josh Raab, David Brown, and Victoria Spencer. I would also like to thank MGMT. for their art direction and beautiful design of the book, especially Alicia Cheng, Sarah Gephart, Sarah Mohammadi, Ian Keliher, Olivia de Salve Villedieu, and Federico Pérez Villoro.

I am particularly grateful to my publisher, Rodale, Inc., and to its talented CEO, Maria Rodale. This is also my third book with Rodale. Their commitment to the environment has been legendary for many years. The Rodale team with whom I worked, including Gail Gonzales, Jennifer Levesque, Yelena Nesbit, Aly Mostel, and Angie Giammarino, did an excellent job. Thank you!

My agent, Andrew Wylie, has, as always, provided extremely valuable guidance and help. Those who are privileged to work with Andrew know that his assistance invariably goes well above and beyond the normal course of duty. Thanks also to Charles Buchan at the Wylie Agency.

This book is a companion to the movie of the same title, *An Inconvenient Sequel: Truth to Power,* which is being released simultaneously. As a result, it has naturally benefitted enormously from the wisdom of the movie's amazing codirectors, Bonni Cohen and Jon Shenk, from its hard-working producers, Diane Weyermann and Richard Berge, and from Sara Dosa and the other talented professionals at Actual Films.

Moreover, none of this would have been possible without the incredible team at Jeff Skoll's Participant Media, including CEO David Linde, the aforementioned Diane Weyermann, Christina Kounelias, and Sam Neswick.

I want to give special personal thanks to Jeff Skoll. Without his encouragement and advice, neither the movie nor the book would have emerged. In addition, I want to thank Lindsey Spindle at the Jeff Skoll Group for the invaluable help that she has provided.

I also want to thank the very able team at Paramount Pictures, headed by Jim Gianopulos. Megan Colligan has played an especially important role. Thanks also to Peter Giannascoli and Katie Martin Kelly.

I also want to thank Mike Feldman, Jason Miner, and Deb Greenspan at the Glover Park Group for all of their help and advice over the past two years on both this book and the companion movie.

My partners and colleagues at Generation Investment Management have been extremely helpful in developing the understandings presented in this book of the Sustainability Revolution, market and investment trends, and the speed with which technologies for renewable energy production, energy storage, electric vehicles, efficiency improvements, and related technologies are developing and spreading worldwide.

Likewise, I am grateful for the help and advice of my partners at Kleiner, Perkins, Caulfield and Byers for their advice. I have also learned a great deal about commitment to sustainability in the marketplace from the men and women at Apple, Inc., with whom it has been my privilege to work for many years.

(In the interest of full disclosure, I have a small indirect investment in two of the companies mentioned in the text, Proterra and ChargePoint, and larger, direct investments in Apple and Google, both of which are also mentioned in the book.)

The Climate Reality Project, headed by Ken Berlin, has been enormously helpful throughout this project. I am very grateful to the board members, professional staff, donors and supporters, and the more than 12,000 trained Climate Reality Leaders around the world for their inspiration. One hundred percent of my share of the profits from the sale of this book (and from the movie) is being donated to The Climate Reality Project.

Nicklen/National Geographic Creative; p. 238: Courtesy of Oren Lyons; p. 239: AP Photo/Craig Ruttle; p. 240: Chris Ratcliffe/Bloomberg/Getty Images; p. 243: Dave and Les Jacobs/Getty Images; p. 245: Courtesy of Wei-Tai Kwok; p. 247: Reed Kaestner/Getty Images; p. 248: Rasmus Hjortshøj - COAST Studio; p. 253: Joseph Sohm/Shutterstock.com; p. 254: AP Photo/Patrick Semansky; pp. 256–257: Jeffrey Markowitz/Sygma via Getty Images; p. 258: Courtesy of Steven Miles; p. 259: © Brett Monroe Garner; pp. 262–263: Klaus-Dietmar Gabbert/picture-alliance/dpa/AP Images; p.264: Arterra/UIG via Getty Images; p. 267: Artur Debat/Getty Images; p. 268: Cynthia van Elk/De Beeldunie; pp. 270–271: Bloomberg/Getty Images; p. 272: Courtesy of TAL Apparel; p. 274: Lucia Griggi/Redux; p. 275: Nike; p. 277: Luisa De Paola/AFP/Getty Images; pp. 278–279: Niklas Halle'n/AFP/Getty Images; p. 281: Daily Camera; p. 284: KK Ottesen; p. 285: Melissa Bender Photography; pp. 286–287: © Greenpeace; pp. 290–291: Jon Furlong/jonfurlong.com; p. 292 (clockwise from top left): Sandy Huffaker/AFP/Getty Images, Courtesy of Paramount, Sandy Huffaker/AFP/Getty Images, Paul Abernethy/Global Citizen, Michael Nigro/Pacific Press/LightRocket via Getty Images, Saul Loeb/AFP/Getty Images; p. 293 (top): John MacDougall/AFP/Getty Images, (bottom): Illustration courtesy of Shepard Fairey/obeygiant.com; p. 296: Courtesy of The Climate Reality Project; pp. 300–301: Courtesy of The Climate Reality Project; p. 304 (top): NASA; p. 306 (from top): DarrenRD CC BY-SA 4.0, Dirk Ercken/Shutterstock.com, © SevArt/Pond5; p. 309: Terray Sylvester; p. 310: Azuri Technologies Ltd.; p. 313: Reuters/Adrees Latif; pp. 314–315 (first row, from left): Courtesy of Lina Carrascal, Rafsanul Hoque, Kara Jess Rondina, Courtesy of Rituraj "Raj" Phukan, Michael Dawson, Courtesy of Sankalp Mohan Sharma, Fundación Global Democracia y Desarrollo, FUNGLODE, Courtesy of Lina Carrascal, Sahin Kumar Lohani; (second row, from left): Courtesy of Rituraj "Raj" Phukan, Courtesy of Nana Firman, Thai PBS, Jennifer Smith, Rehia Qais; (third row, from left): Courtesy of Nana Firman, Courtesy of Nora J. Coker, Courtesy of Guilherme Sortino, John Leo Algo, Joe Douthwright; (fourth row, from left): Courtesy of Eric Novak, CREO-ANTOFAGASTA, Courtesy of Jess Reese, Jyoti Bhandari, Courtesy of Rohit Prakash; (fifth row, from left): Courtesy of Mehak Masood, Dan Brannan EdGlenToday.com, Courtesy of Duncan Noble, Fundación Éforo; (sixth row, from left): Courtesy of Rituraj "Raj" Phukan, Clearwater Bay School, Mean Mustard of Cebu City, Courtesy of Sankalp Mohan Sharma, Marc R. Caratao, Julianne Reynolds, Bruce Bekkar; (seventh row, from left): Jay Wilson, Adopt a Native Elder; (eighth row, from left): Explora Science Museum, Courtesy of Brian Ettling, IEEE SEECS, Rehia Qais, Courtesy of Andrea McGimsey; (ninth row, from left): JP Santos, SEREMI MINVU LOS LAGOS, Courtesy of CB Ramkumar, WESA 2017, Courtesy of Rashida Atthar, Nat Giambalvo, Courtesy of Allison Arteaga

Thank you to these Climate Reality Leaders featured on pp. 314–315: Alice Giambalvo, Allison Arteaga, Andrea McGimsey, Brian Ettling, Bruce Bekkar, CB Ramkumar, Claudia Cedano Belliard, Delaney Reynolds, Duncan Noble, Eden Vitoff, Eric Novak, Erica Largen, Guilherme Sortino, Hernan Silva, Jacqueline Lucero, Janice Kirsch, Javaria Qais Joiya, Jess Reese, John Leo Algo, Katarina Hazuchova, Lina Carrascal, Mara Cantonao, Martin Rabbia, Mehak Masood, Michele Douglas Eleta, Nana Firman, Nora Foster-Coker, Patricia McArdle, Pragya Ghimire, Raj Phukan, Rashida Atthar, Rohit Prakash, Rubina Karki, Ryan Anthony Bestre, Sankalp Mohan Sharma, Stephen Bieda, Stuart Scott, Syeda Nishat Naila, Umay Salma, Viveik Saigal, and Wei-Tai Kwok.

Text on p. 196 courtesy of The Greenville News.

This book was produced by

MELCHER MEDIA

124 West 13th Street, New York, NY 10011
melcher.com

President and CEO: Charles Melcher
Vice President and COO: Bonnie Eldon
Executive Editor/Producer: Lauren Nathan
Production Director: Susan Lynch
Editor/Producer: Josh Raab
Assistant Editor/Producer: Victoria Spencer
Consulting Editor: David E. Brown

Designed by MGMT. design
Illustrations by Danaiphan Washareewongse
Information graphics by MGMT. design

Melcher Media would like to thank:
Mary Bakija, Callie Barlow, Jess Bass, Emma
Blackwood, Tova Carlin, Amelie Cherlin, Jon
Christian, Karl Daum, Michel Diniz de Carvalho,
Shannon Fanuko, Barbara Gogan, Ashita Gona,
Evan Greer, Mary Hart, Luke Jarvis, Aaron Kenedi,
Karolina Manko, Emma McIntosh, Kate Osba,
Nola Romano, Rachel Schlotfeldt, Daisy Simmons,
Michelle Wolfe, Megan Worman, Katy Yudin, and
Gabe Zetter.